You Must Be *Very* Intelligent

Karin Bodewits

You Must Be *Very* Intelligent

The PhD Delusion

Karin Bodewits
Munich, Germany

Edited by Stewart Hennessey

ISBN 978-3-319-59320-3   ISBN 978-3-319-59321-0 (eBook)
DOI 10.1007/978-3-319-59321-0

Library of Congress Control Number: 2017945402

Illustrated by Vanessa Czerwenka
Book cover design by www.unionstudio.co.uk

Printed on acid-free paper

This Springer imprint is published by Springer Nature
The registered company is Springer International Publishing AG
The registered company address is: Gewerbestrasse 11, 6330 Cham, Switzerland

I dedicate this book to all the proud parents of PhD students.
Keep your spirits up…
… and please don't read this book.

Preface

Ever since I finished my PhD, or maybe even before that point, I knew I had to write this book. However, this is not a diary of my time as a PhD student at the University of Edinburgh. Having said that, it is not quite a work of all-fiction either. If it were, I would probably have added some Frankenstein-like personalities and mad, sleep-deprived, eccentric geniuses, as beloved by atmospheric old movies. Or I would have described secretive, non-ethical research taking place in dank basements beneath cloisters, proving that scientists are amoral psychopaths. I did meet some people I could imagine creating a three-headed sheep for shits and giggles but I never actually saw anyone trying it.

However, I saw stuff that was dramatically dark, barking mad and hilariously ridiculous, but in an everyday way. I saw the monsters beneath the meniscus of human nature surfacing in a supposedly sedate world; of frustrated egos the size of Africa, where competition is pathological, volcanic rages seethe and tin pot dictators are drunk on oh-such petty power. It's a world where glory is the goal and desperation is the order of the day; a world where young adults are forced into roles that make *Lord of the Flies* look like Enid Blyton.

It was an education. And it taught me to be wary of education.

Karin Bodewits

Prologue

The evening sky is vivid orange as if the sun is leaking colour into it. In a few minutes this warmth will disappear and the city of Edinburgh will fall into darkness. From the air I will be able to sight only lines of streetlights, Edinburgh Castle and the Forth Road Bridge. And then the plane will remove me from this island which I have come to love. When this happens I let out a deep sigh, thinking "It's over. It is really over…"

The backpack, between my feet, is heavy from the thesis and copious notes I scribbled in preparation for my PhD defence. If I could open the tiny window next to me in the airplane cabin I would joyously throw it all into the night sky.

My stomach feels queasy, the clothes I am wearing smell of puke, I smell of puke. It was only a few hours ago that I stood in front of the synagogue on Salisbury Road and threw up the alcohol still in my stomach from the evening before. I'm not anti-Semitic, I just happened to be there when I couldn't keep it in any more. Friend Felix had stayed with me the whole night, to make sure I would survive. I have no memory of his presence in my room last night but he was sitting at my bedside this morning when I awoke in a pool of shame. When I noticed the towels on my bathroom floor I understood why he had stayed – I had indeed needed help.

During the days leading up to my exam I hardly ate, owing to nerves. I only had half a pint of Guinness the evening before to calm me down and help me sleep. After the exam I still wasn't hungry. I went straight to the pub, first on campus, then in town… I danced and drank… mainly drank.

Almost four years earlier I had started my PhD at the University of Edinburgh like a good girl. At that time I truly believed that researchers in the ivory tower were idealists driven by the desire to make the world a better place through the advancement of knowledge. My faith in the university system, as the crucible of meritocratic refinement, was absolute: only the extremely knowledgeable and wonderfully intelligent would ever hold chairs and professorships. If you are not smart enough you would have to leave. Hence, like many other PhD students, I arrived at a high-ranked university feeling excited and privileged, brimming with hope and wondering if I could prove myself worthy. I wasn't only an ambitious student, and an aspiring scientist, but also a beloved daughter and sister. I had been dating a sweet guy called Daniel for years; he would follow me to Edinburgh where we would meet our inevitable demise. I believed in the goodness of people, I believed in the goodness of the ivory tower and I saw professors as intellectual role models. I would follow their path, as far as I could…

That seems like another person in another life. When I started my PhD I had been scared I wouldn't be clever enough to become a doctor, but I did believe I had at least done enough to prepare myself for starting this degree. The truth is, I was clever enough but nothing had prepared me. Looking back, whilst cringing in embarrassment at my naiveté, I wonder now if I was actually unlucky to be accepted?

I now have my title, the two letters I have been craving, I have made my parents proud – I am a doctor… What does that mean? Marginally less than a Girl Guides' Camping Badge. This is what I've learned.

Trembling and hazy, I gaze down at my lap where there is a bread roll, which I promised Felix I would eat. My hands are still pale and shaky, and my legs throb with exhaustion. What course my life will now run, I know not. I only know it will be a life outside of the lab, away from university, away from the academic path and, crucially, very far away from my PhD supervisor, Mark. Despite my physical condition, this is like a euphoric dream. Listening to Dylan's *Blowin' in the Wind*, I look out of the window into the sea, I am away, forever, it really is over…

Contents

Part I: Before

Chapter 1

I wake to the sound of raindrops tapping on the window of the tiny room in Pollock Halls of Residence. Situated on the edge of Holyrood Park near the foot of the dramatic little mountain, Arthur's Seat, this should be a sweet awakening. But I'm soaked in feverish sweat, my stomach is cramping crazily with every little movement, my head is pounding like a gong and my back feels like I've ridden a camel across cobbles all night.

The mattress I am lying on is hanging through the bed frame as if something weird and dramatic has recently occurred here, such as two sumo wrestlers having torrid sex. It resembles lying in a toaster rather than a bed. The wooden desk, attached to the window sill, looks askew on just one twisted hinge. I am not sure if it really is, or if my vision is distorted by the weird bed. From the outside, Pollock Halls is a beautiful and renowned student residency, which is rented out during semester breaks as a B&B. From the inside, it's like a slapdash sitcom set – at least the room I'm staying in.

For a few minutes I lie in silence, wondering how I will ever get through this day. In three hours I am expected at King's Buildings, the university campus where the School of Chemistry is located, to interview for a PhD position. Since coming back to Europe one week ago, after almost a year working for Unilever in China, I had been at my parents' place in the North East of the Netherlands. In the last seven years, I hadn't spent more than the occasional night in the village where I grew up, but for now I had nowhere else to live. My childhood bed had been binned long ago and my room converted into a guest room. I didn't want to stay in the village for a

© Springer International Publishing AG 2017
K. Bodewits, *You Must Be* Very *Intelligent*,
DOI 10.1007/978-3-319-59321-0_1

week, but I had been ill. At first I thought I was suffering a prolonged hang-over from my goodbye-to-Shanghai night out. When the hangover hung around all evening and even seemed to intensify with ibuprofen and coffee, I wondered if I had perhaps caught a virus. It was not until I was pretty much hallucinating and loosing litres of sweat that I started to fear a strange Asian parasite was flourishing within me.

I had postponed my flight to Edinburgh for a couple of days, cutting out the tourist part of my trip. The sickness had subsided to a flu-like condition, but it had drained me; I feel exhausted and defeated. And I want to collapse, for days. But cancelling such an important appointment was never an option. I am not sure if postponing would have diminished my chances of getting a PhD but I didn't want to risk it. Now there was nothing for it but to haul my woozy body out of the godforsaken bed and pretend to be senti-ent for the duration of an interview. I convince myself that painkillers and fever blockers will get me through.

I applied for this PhD position – conducting research on cystic fibrosis – just a few weeks ago, after reading about it on a job portal for scientists. In fact I don't really have any particular interest in cystic fibrosis but during the previous years I had learned that it doesn't really matter to me what I work on. It could just as easily have been Alzheimer's or cancer or some strange single cell organism no one really gives a damn about. Of course there are scientists obsessed with finding a treatment for one particular dis-ease. But I am not. I am not sick myself, and nobody close to me suffers from a strange illness to get emotionally attached to, or even a mainstream disease to get particularly zealous about. In the area that I graduated you zoom in tight on the tiniest details, like single proteins or molecules. The relevant illness is very distant and abstract. However, the larger picture is important in order to avoid social isolation at parties. Being a cancer researcher draws interest and respect. Studying complex fluids in an inexplic-able matrix that no one has ever heard of is everything but glamorous and makes you an intellectual freak.

For me, however, the main appeal is contributing to fundamental know-how that might find a future application. It was only after I spent time in a couple of industries, and had several unconvincing job interviews, that I decided my future would be in science, nowhere else. I had never been more than a trumped-up intern in the "real world" though I enjoyed many aspects of it: I found it extremely cool to sit on the 23rd floor of a skyscraper in Shanghai and video-interview with the Boston Consulting Group in

Amsterdam. I felt important and wanted when Unilever paid for someone to chauffeur me around town. On top of that they provided a salary and spoiled me with a ludicrously large and luxurious flat in Shanghai. But I didn't belong in that world. I don't care about optimising returns, extensive and allegedly important meetings, developing shampoos that make your hair shine EVEN more (oh wow, how interesting, well worth several years study…) or producing milk with additives no one really needs. I hate cutlery etiquette. And I hate people talking for the sake of talking. I don't want to spend my holidays skiing with colleagues, pretending to be part of a happy, smug family. I don't want to smile for the sake of the customer. I'm not a team player. I am an individualist, content to sit in an office working for hours upon hours alone. I am a scientist…

When opting for academia, I fully understood that from a financial and quality-of-life perspective I was taking a step back. From living like a queen in China, I would go to mashed potatoes and peas. A Unilever secretary would book me business class flights. If I'm lucky Dr. McLean might cover a self-booked EasyJet ticket. But if it means I can advance science and do what I love, so be it.

The position is at a famous and prestigious university – good credentials for the future. And as the cherry on the cake, it means working in a country I have never been to, yet am keen to live in. When I sent my CV and cover letter, the last thing I was expecting was an invitation to present myself. I graduated with good marks, but the competition at this university, ranked in the top ten in Europe, would be fierce. Plus, my background wasn't exactly what they were looking for. I had been thrilled when the email came in from Dr. McLean, my potential PhD supervisor, telling me I had made it through initial selection and would be welcome to fly to Scotland for an interview. And here I am.

I slowly get up from the toaster bed, hearing my spine crack and reposition. I have to concentrate all my energy on not throwing up. Slowly I bend over to my backpack and get ibuprofen, a diarrhoea blocker and Vomex. I'm not sure if I'm supposed to take these drugs together but I feel too weak to open my laptop and research the matter. *Why would I care? It couldn't get worse?* After swallowing the three meds I set my alarm for an hour later.

There is no way my back will accept another nap in the bread toaster position, so I take the sheets off the mattress, resplendent with old sweat circles, and carefully lay them on the tiny bit of floor space between the bed and the wall. It is hard and cramped but comparatively luxurious.

When the alarm goes off, and I believe I've only watched the roof of the room for the last hour, I feel better. I down an old bottle of Coke; it's flat but I get the caffeine rush. I take clothes out of my backpack which my mum had packed while I was crashed sick on her couch. She assembled and ironed an outfit for the interview based on wishes I had muttered, rather deliriously I now suspect. I put them on and conclude that the ensemble doesn't work; light brown trousers and a cream-coloured top – neither are items I would ever be likely to wear and they don't go together. Admittedly, they do come from my stack of clothes, and I'm not sure what clothes would best present who I am anyway. But I hope it's not this prim, dull clobber…

I feel a weak spark of adrenaline going through my body, like a vibrator on its final stripe of battery. I stand in front of the mirror stuck on the cheap built-in cabinet, and conclude that I look awful, sort of Kate Middleton's junkie cousin trying to look posh. My face is pallid; even the freckles have faded to grey, and I've got comically puffy eyes. *He is not going to take me. I look much more like a lost extra from The Walking Dead than a ground-breaking research scientist…*

I turn away from the mirror and tell myself to concentrate on my inter-view instead. Worries shoot through my head about the presentation I have to give in two hours. I have managed to put some PowerPoint slides together, about my previous experiences and research projects, but I didn't really get round to the words part. I need something to accompany the smile and slides, the "Hello, I'm great" show. I open my laptop, call up the PowerPoint presentation and start to talk. It's terrible, I think, I presume, I don't actually know, my head is spinning…

In my hazy state I decide I have to refine my detailed, scientific presenta-tion whilst walking to campus, though I don't even know the way. I check on Google maps and have a last glance in the mirror at my mismatching clothes atop the mismatching shoes I just added. *Yip, I look like prime shit.* I pull my hair back into a ponytail, a last touch of general awfulness, and walk out of the door feeling visually offensive.

It's merely drizzling now. The cold drops on my face are enlivening, the adrenalin is going up a bit and the dizziness I have been feeling for days is suddenly on the backburner. It's almost half an hour's walk – with an extra small loop because I'm lost – and I believe I have rehearsed my presentation twice. I arrive at the old building with a modern sign in front of it: "School of Chemistry."

I open the thick wooden door and step into the entrance hall, which looks like an old train station with a curiously undecorated roof. Beautiful memorial plants with thick, wooden stems climb up to the first floor. On the left side is a counter for a receptionist, but the seat is empty. The big modern flat screen in the corner – announcing the school's programme – draws the eye, mainly because it is horribly gaudy.

There is a long wide corridor leading all the way to the back of the large building. On one side of the corridor there are lecture theatres and a coffee room with couches – a space called *The Museum*. I vaguely wonder from whence that name derives because cheap cotton couches with spindly legs hardly constitute exhibits.

Though I printed out a plan of the school, and even marked the right office with a pen, I struggle to orientate myself. The dizziness is returning and I am devoured by the desire to lie down and sleep, maybe on one of those cheap cotton couches, anywhere really… Quickly, I follow a sign to a toilet where I get two 400 mg tablets of ibuprofen out of my bag and swallow them.

"Are you okay?" enquires a short girl with gorgeous gold curly hair.

She is washing her hands at the sink next to me. "Yes, just a bit sick, that's all," I say, forcing a smile which possibly looks rather eerie, from a ghost.

"Can I do something for you?" she asks, with a look of deep concern.

"Do you know where office 221 is?"

She looks nonplussed: "Who are you looking for?"

"Doctor McLean."

She points to the left. "Take the stairs, turn right, first on the left and it's one of the offices in that corridor."

"I'll find that."

"Are you here for your PhD interview?"

"Yes."

"In the McLean group?"

"Yes."

She nods as if this has satisfied all the interest she might have had in my predicament. While opening the door, she says: "Apparently, he is not the easiest to work for."

My antennae pick this up but before I can ingest it and formulate a question, she leaves me at the sink. *Who needs it easy? It doesn't matter for now, I'm here anyway…*

I follow her directions, nerves rising within me. Hesitantly I knock on the door and a loud voice beckons me forth. With the enthusiasm of someone who has just placed a red hot pepper between his arse cheeks, Dr. McLean is jumping up from his chair to introduce himself. His outstretched hand is extremely masculine and the handshake suitably vigorous. "Mark," he says.

His eyes are large, and they radiate energy. Of course I googled him and he is indeed as young as he looks – late thirties I think. He's wearing jeans, sneakers and a simple shirt, all quite fashionable and fitting his age.

He enquires if I had a good trip and how I like Edinburgh so far?

"Great!" I lie, rather smoothly I think.

In fact I haven't seen anything of Edinburgh. I took the bus from the air-port to the main train station then a taxi to Pollock Halls. But the last thing I want to admit is that I have been too sick to travel here earlier. I don't want to look feeble. I want to look like an alpha female, whatever that is...

He takes a keychain from his desk and grabs a laptop. "Let's go to the lab straight away!"

He strides through the doorway and along the corridor at a merry clip. I struggle to keep up. He halts in front of a lab door numbered 262, in remarkably large digits, opens it very quickly and just as speedily shows me around the three small rooms within. I don't really take anything in except that most of the equipment is old and the lab is very messy. The contrast is striking: coming from the new, ultra-controlled and clean Unilever R&D labs with super-expensive equipment to a lab in an old academic building, run by a young group leader clearly not floating money. I swallow hard and try to hide my faint alarm. Mark may have sensed my disappointment. "It's a bit messy here, but don't worry, we're in the process of cleaning it."

He introduces me to the few people, all around my age, working on the benches. I can't remember their names, if they even registered at all; my brain feels as if it is being run by a virus rather than by me. Mark makes a few jokes with the people in the lab, which only an insider would under-stand, but it doesn't really persuade me that everything is chummy here. The atmosphere is somehow guarded. Mark then loudly announces I will be giving a presentation in five minutes which they are all to attend.

He isn't intimidating as I had expected after the comment from the girl in the toilet. He talks a lot and doesn't ask me many questions about myself. I am vaguely aware that normally this would feel odd but today it feels like a blessing.

He guides me to a small seminar room down the corridor and opens the laptop he has been carrying.

"Please upload your presentation; I'll just fetch the projector from the reception."

I can't recall entering a seminar room in recent years where a projector was not part of the everyday furniture. Apart from a few tables and chairs, and a blackboard with a dried-up sponge lying next to it, this room is empty.

"Ideally I am looking for a chemist, and you are a biologist," he says when he returns with the projector.

"I like chemistry," I say, while transferring my presentation from the USB to the laptop. "I am looking forward to learning more of it."

"Good, your background seems impressive enough. You will learn the chemistry."

There is a knock on the door and an old man enters.

"Hi, great you could make it," Mark says enthusiastically, shaking the man's hand. "This is Karin, all the way from the Netherlands. Karin, this is James. He will be co-supervising the project and he is THE expert when it comes to cystic fibrosis. We've been collaborating for years now."

"Hi Karin, nice to meet you," James says, offering a warm hand.

He exudes an air of old school reassurance; partly just by wearing his long service very visibly in his demeanour. He is grey-haired, with a round face and grey, watery eyes; a biologist who somehow fulfills all the stereotypes of a biologist of a bygone generation.

"I won't be able to stay long I'm afraid. I'm expected back in the hospital in less than an hour. I'm sure you guys will be fine without me."

His voice is less energetic than Mark's, but it sounds confident and calm. "Yup," Mark says.

Six other people filter into the room, mainly from the benches of the lab I think. Mark introduces me with a few friendly words and invites me to start my presentation. I stand up and tell them about my previous research projects, in industry and academia. Due to confidentiality agreements I had to sign in industry there are not many details I can share about what I did exactly. But it doesn't matter. I feel much more confident than I felt two hours ago, and for a moment it seems that there is no virus living in my body.

After my presentation, Mark invites the audience to ask questions. James has a clarifying question about the research on dandruff I have conducted, which is easy to answer. With the second call for questions, they all peer at me but don't ask anything. After a few seconds the silence is awkward, but more grim seconds pass before Mark finally thanks everyone for their attendance and calls the end of the meeting. That was weird, not encouraging.

"Great presentation, Karin," James says. "I don't want to be rude, but I have to make myself scarce now."

"No worries, it was a pleasure to meet you," I say, forcing a smile.

James talks to Mark briefly about a project they are working on and leaves the room. "Nice work," says Mark, stuffing the projector back into the black bag.

We drop the projector back at administration and walk to his office passing an open toilet door along the way; I glimpse myself in the mirror. *I still look really shit.*

I sit down on a chair in his office and Mark starts talking to me about the project he has in mind. "I've just been talking to Mike Wood in Canada, we're really on the right track to publish everything about this pathway… No it isn't easy, but we're getting there… privilege to work with Mike

Wood, excellent scientist, a wonderful person. I visited him last year with my fiancé on our holidays in Canada… You could go to his lab! You can learn so many different techniques… We can explore KdtA… very interesting protein… not only one glucose to Lipid A, but actually two, very strange mechanism, almost like a polymerase… and all those proteins that lie in the inner membrane, it can explore all of them… we have an excellent postdoc here in the hospital as well, Brian, he can help you with all your research…"

He is a lively and enthusiastic person but he talks with a strong Irish accent so it took me at least ten minutes to get to grips with what he was saying. And by then I had entirely lost the general thrust. He rambled on for over an hour about things I did not really understand. I noticed he talked about people as if they were the most famous scientists in the world, and looks surprised when he sees I have no idea who they are. But it doesn't seem to matter. He throws in names of people in his lab as if I would know who they are and what they are working on. He shows me incomprehensible paper after incomprehensible paper and – thank God – does not really bother asking me anything. By now I could feel my medication losing its power, though I had recharged by secretly swallowing a new load of pills after the presentation. My eyelids have never felt so heavy and I'm scared they will fall shut but – hallelujah! – Mark finally wraps up.

"Hanna will take you for lunch," he says.

Oh no! I desperately want to curl up and sleep for a week. I need to go back to bed. And who the hell is Hanna anyway?

Mark had mentioned her name several times during the last hour, and apparently her PhD project is similar to mine, not that I understand either yet. Probably Hanna had been one of the girls sitting in the presentation, one of the names I failed to register.

Mark walks me to the lab and indicates to a black-haired, brown-eyed girl with thick red glasses that it is time to take me for lunch. She has a soft expression on her beautiful girlish face. She looks innocent somehow. Mark shakes my hand with the words, "You will start the first week of September. I very much look forward to working with you. Look for an apartment in Leith, it's a nice part of town and not too expensive. Give me a shout if you need any help."

He gives Hanna money to buy lunch for us, says goodbye and walks back to his office. *Does that mean he offered me the job? Without officially offering it to me? He seems to assume that I will be taking it, I suppose he's right, I think, I can't think…*

Hanna and I walk out of the Chemistry Department and cross campus. She points at a two-storey building opposite of us. "That is the canteen we normally go to, but today we'll go to the Darwin Building canteen. It has a lovely view."

She tells me what is being hosted in the different buildings we pass. "It is all old here, but you get used to that. The only thing you won't get used to is the darkness in winter and the food."

Hanna tells me she is about to start the last year of her PhD, which usually takes a total of three years in the UK. She did her undergraduate biology degree here as well, though she is Norwegian. "I'm a biologist, not a chemist and, to be honest, I never wanted to work for Mark. I wanted to do my PhD with James Ainsley. But James is nearly retired, and does not supervise PhD students anymore, so I ended up with Mark. You will like James very much."

"That's odd, because I think Mark told me that if I got this position, James Ainsley was going to be my second supervisor and would be very much involved in the project."

"No worries. He is one of those professors who will never retire, just officially."

At the University of Groningen, where I got my undergraduate degree, we had a couple of professors who were both well over eighty. They could barely walk, but they would come to the university every day. They shared an office on the second floor next to the teaching labs. No one knew what they were doing but clearly they were devoted to science. I liked seeing them, slowly making their way through the long corridor, being overtaken by students who were less than a quarter of their age. I loved their expression of passion, the fact that they gave their lives to science. *What an experience it would be to work for one of them.*

"Are you going to accept the position?" Hanna asks, in the elevator.

She just pressed the button to go to the seventh floor; I feel trapped and bound to answer.

"I'm not sure. I'm a bit confused. It just sounded to me that he offered it to me, but then he wrote a few days ago that there are other candidates as well."

"No, there are no other candidates. He might have received more applications but I haven't seen anyone else here."

"Would be difficult to say no, it's an amazing opportunity."

She rolls her eyes, but avoids eye contact as if she was about to tell me something. Eventually she speaks, "I don't know about that," she says cautiously. Then she pauses for a few seconds. "But you will be very lucky to have James as your second supervisor. The other students don't have him."

I'm awake enough to pick up the ominous note but not awake enough to care terribly much.

The canteen has a beautiful view over Edinburgh but food-wise it is horrific; a few unappealing salads and muffins seem to be all there is on offer. Maybe it's because it is well past lunchtime. I'm not hungry anyway and opt for a coffee. After a few large gulps of the black liquid which, at best, is only analogous to coffee, I feel newly energised. We chitchat about Norway, the UK and the Netherlands. It's a nice and friendly conversation but, thankfully, Hanna keeps it short. She needs to go back to the lab and I need to go to bed or at least to the sheets on the floor. She shakes my hand before the chemistry building and closes with: "Whatever you decide, don't move to Leith. You really don't want to be there. It's a shithole!"

I head back to Pollock Halls, exhausted, legs wobbling as if I'm drunk. By the time I arrive I feel weirdly dehydrated, so I drink three large glasses of water before tumbling to the floor and into sleep, in seconds.

When I wake it is still light outside, but clearly evening. My face has some colour again though my hair is a comical mess. I am hungry. I haven't eaten since the banana I forced into myself just before the interview, at ten this morning, and my phone tells me it is just after 9:00 p.m. I get into a pair of jeans and head out into the street where it is still raining. I ask a guy, roughly my age, for directions to a takeaway. He points me to a local kebab shop. I thank him and give him a friendly smile. He smiles back and for a short moment I feel the physical attraction between us, like two magnets coming too close. *Thank God, my body is functioning again.* The moment passes and we set off in opposite directions. Just before reaching the traffic lights I need to cross, the guy taps on my shoulder and hands me a piece of paper with his telephone number. "Give me a ring if you fancy." I take the paper from him and give him another smile. "I might do," feeling kind of guilty that I didn't say straight away I have a boyfriend and I know the chances of me ever phoning him are infinitesimal. And yet, one day in the unknowable future I will phone this stranger, and not by accident.

I slip the paper into my jeans pocket. The guy heads off as quickly as he arrived and I'm crossing to the other side of the street.

I walk to the kebab shop broadcasting caterwauling music in an unknown exotic language. There is a guy behind the counter in shorts and a polo shirt that sits tight around his biceps. I order myself a chicken wrap with tomatoes, salad and garlic sauce, and take a pack of cigarettes from the counter and a can of coconut milk from the fridge. All of it I take back to Pollock Halls where I sit in front of the building under a shelter.

It seems deserted here, and just at the moment I finish my wrap and start to believe I am the only guest in this residence, a grey-haired guy joins me under the shelter. He's got bowlegs, a bent back, and a hawk nose. We both light a cigarette, and just to punctuate the silence I ask what brought him here. He tells me that he is a medical biologist, a Professor from Israel, and has just attended a conference in Glasgow, an hour away. He wanted to see Edinburgh before flying back to his home country. We talk a bit about everything and nothing – his research, my PhD interview, his wife and his daughters, his first grandchild, my boyfriend. It is a pleasant enough conversation. He fetches a couple of beers from his room that we drink under the shelter while we smoke another cigarette. Then another two…

It is almost midnight when I finally stand up.

I say, "I very much enjoyed talking to you, but my bed is calling. I still want to see a bit of Edinburgh before my flight tomorrow."

"You want to join me in my room?" he asks.

Knowing the size of my room, which is basically not more than two walls with a tiny desk and a bed in between, I'm pretty sure that I know what he wants to do in there. *Does this guy, who is even older than my dad, and who has just told me he recently got his first grandchild, really believe that I would want to have sex with him?* I stare at him in disbelief.

"Are you proposing to have intercourse?" I ask with wide open eyes.

"Yes, that is what I'm proposing."

"I don't want to have sex with you!" I say very loudly and, I suspect, looking disgusted.

He swallows, and seems disappointed. "Some girls say yes."

"Are you joking?" I ask, genuinely stunned and truly unable to imagine anyone wanting to have sex with Quasimodo.

He looks surprised. "What about a blow job?" he asks

"No," I say very clearly, wondering if his dick is as warped as his nose, and his scrotum as crumpled as his forehead… but feeling far from keen to find out.

"Too bad. Well, I had nothing to lose."

"What about dignity?"

I had felt good about the young guy giving me his number. And now Quasimodo says I am actually in his league. It's always yin and yang with sexual self-esteem. But, hey-ho, the signals are lively in Edinburgh…

Chapter 2

"So, do you have to say now if you want to take it?" Daniel asks, handing me an ibuprofen and an espresso.

He sits down next to me on the uncomfortable sofa in his parents' living room – beautifully designed and expensive, but useless.

"It seems he just presumes I will take it."

"You sure you want it?"

"How could I not? It's an excellent opportunity, isn't it?"

"It is a good university. BUT, they are chemists… all freaks."

"The research group looked normal to me."

"The rest of the department won't be!"

"I haven't seen them, but it will be fine. Plus, I will spend lots of my time at the hospital labs."

He is struggling to be happy for me. We have been dating since pretty much the start of our biology studies. We were in the same year but in contrast to me Daniel hasn't finished his master's yet – not even close to it. I've always been quicker than him, had higher marks in every topic, and my internships were fancier than his. He never speaks about it but clearly it frustrates him to always be second – out of two. Plus, I knew by now what I wanted to do with my life. I had a clear goal in mind. I wanted to be a scientist. Daniel, however, was still trying to figure out what to do with his life and how he could make a difference, or even just find some motivation.

I give him a hug and let my hand slide under his T-shirt.

"It will be fine," I whisper in his ear.

"You've got a fever, Ka. Just lie in bed."

"I'm sick of lying around," I lie.

© Springer International Publishing AG 2017
K. Bodewits, *You Must Be Very Intelligent*,
DOI 10.1007/978-3-319-59321-0_2

I would love to lie down but I know that as soon as my body is touching a mattress, Daniel takes the opportunity to lie down next to me and do nothing for the next few hours, or days. The few times I had been sick during our relationship have all, apparently, been good reasons for Daniel to only oxidise as well. It is so dispiriting that I prefer to stay up myself.

My head is soon spinning and I ask Daniel to get me some more coffee and ibuprofen.

On his return, I ask him how his week has been; staying such a long time with his parents, which he hasn't done for years. He had also been in Shanghai, spending some time at the university there, and we returned to Europe together. Our ways had separated at the airport. He had gone to his family in the South and I went to the North. I know his parents' house is not the most uplifting place to be. Their lack of interest in Daniel is quite a phenomenon to behold. Just when he wants to answer my question we hear the keys turn in the lock of the front door. It is either his mum, dad or the cleaner. The footsteps on the wooden floor in the hall sound like a woman's heels – his mum presumably.

Daniel comes close and whispers right into my ear. "Moving with you to Edinburgh sounds very appealing to me right now."

Daniel comes from a much more urbane and sophisticated family than I do. I was raised in a small village in the economically dead North-East of the Netherlands. My parents have a middle class income and belong, along with the farmers, to the upper echelons of local society. But compared to Daniel's family they are rustic people with simple lives. Daniel had been raised in a mansion in a central location of a city in the south of the Netherlands. His parents studied, mine didn't. My parents were careful with money, whereas his were throwing heritage overboard. There had been loads of money in his family for over a century, moving from one generation to the next. His parents still got a fair share of the wealth, although Daniel's grandfather blasted most of it. Apparently there would not be anything left for Daniel and his two younger sisters.

My parents acquired their possessions by way of hard graft. My grandfather from my mum's side was never able to buy the rented farm he ran his whole life and never possessed anything much beyond a small house and a car. My other grandparents lived in a heap of rubble that had been declared "unliveable" by local government before my dad was even born. There was nothing to inherit from my family, at least not financially. My parents were very down to earth, not flamboyant, quite hardy, perhaps a little cold. His parents were flower power hippy dilettantes until well into the 1980s. His mum claimed to have carried the Red Book during her student days but nothing today suggested a communist mind-set. They were capitalist consumers indulging whims with every Euro they could spend. The only thing that my and his parents had in common was that neither had any doctorates in the family – I could be the first.

Of course our roots made Daniel and me different. Since the age of 14, I had side jobs to pay for my studies and other expenses. From peeling flower bulbs to working in a kebab shop or call centre, and later as a teaching assistant, I have seen many things. Daniel had never worked for anything in his life; his parents, so far, had paid for everything, including backpacking trips. A quick call home was all it took to top up his bank account. Daniel knew *Esquire* etiquette, I did not know it even existed. I had spent my youth catching frogs and climbing trees, while he visited museums and cultural festivals. I saw the Alps, he saw the world. He is a dreamer and a materialist envisaging wealth. I am practical and don't give a damn about money.

"Hi Karin, how are you?" his mum asks.

Though most people close to me call me Ka, she is still, after years, calling me by my full name.

She hasn't taken her coat off yet and doesn't look at me, just sort of senses me, while she rips open some envelopes she grabbed from the mailbox on the way in. She only really looks at me when we happen to smoke a cigarette together under the cooker hood.

"Fine yes, just came back from Edinburgh."

"How was the city?" Daniel asks as his mother doesn't react.

"Didn't see much of it, went on one of those 'hop on-hop off' buses this morning to catch a little."

Daniel is trying not to smirk. He knows how deathly I must have felt jumping onto one of those buses. We've been dating for so long that we can link precise emotions to every facial expression and vocal inflection. We are not only partners but also, over the years, we have become friends – inseparable buddies. To me it became kind of boring to know each other that well, and I couldn't help desiring a fresh, exciting new partner. Ideally I could keep Daniel as my comfort zone, and have a new partner on top. I know this is not possible, unless I live a constant lie, and I am a little perturbed to be having such fantasies at all.

"From what I saw it is a lovely town," I add.

"What did you do there again?"

Her eyes are unwaveringly focused on a letter while she talks to us.

"I interviewed for a PhD position at the University of Edinburgh."

She finally glances up, in our direction.

"Did you get it?"

"Seems like it, yes. I just need to decide whether to take it or not."

She smiles at me and sounds satisfied, "We will have a doctor in our family."

Before bed I phone my parents to tell them about the interview and inform them that I had just sent an email to accept the position. My mum sounds excited and I can envisage my dad smiling on the other side of the line. I guess my parents are not only happy that I am going to study at such a famous university but that there is now also a clear end-date to me living with them. They are no longer accustomed to having a daughter in the house, and I am pretty sure the amount of stuff I had spread over their floor, due to the lack of a wardrobe, is bothering them quite deeply. We all knew beforehand that this could only work temporarily.

After the phone call, I lie down. I fall asleep in Daniel's arms. I am dreaming about having sex with a grey-haired guy with a crooked back on the half-pipe mattress. Oh dear.

Part II: Year 1

Chapter 3

I still have time to smoke a cigarette on the platform before the train departs. It is a warm September morning, but I feel shivers rippling over my body. Between the skin on my back and the tank top I am wearing there is a thin layer of cold sweat. I did not have time to shower before leaving and I'm still wearing the same clothes I wore yesterday. I had spent the night hopping from one pub to the next and by the time I got back to my parents' place it was time to wake them up to drive me to the station. I have been smoking all night yet this cigarette is making me sick. I feel out of it: *Am I leaving to become a mature PhD student, in the prestigious University of Edinburgh?*

I nestle into a seat and cover my shaking body with a thick winter coat. Just before Amersfoort I'm stirred awake to see the train conductor wearing an expression which suggests it was an annoying effort to rouse me. I hand him my ticket and hope to get it back as soon as possible so I can continue sleeping.

"Young lady, you cannot use a discount ticket before nine in the morning."

I look at my mobile. "But it's 8:45 a.m..." I mumble and try to look coy.

"I can either give you a fine of thirty-five euros, or you get out in Amersfoort and wait for the next train."

"In Amersfoort?" I whisper in disbelief.

"What's wrong with Amersfoort?"

"I'm not sure, it just never occurred to me to get out there."

He shakes his head and walks away.

The wait in Amersfoort seems to be the longest twenty minutes of my life. I feel even colder than before and my eyes are heavy. I smoke another cigarette

© Springer International Publishing AG 2017
K. Bodewits, *You Must Be Very Intelligent*,
DOI 10.1007/978-3-319-59321-0_3

amid the other addicts, all corralled around a smoker's pole in a little cluster of shame. I know I must be swathed in odours of alcohol, sweat and cigarettes. If I were more awake I might find the grace to be embarrassed.

My mind wanders to yesterday evening. I am overwhelmed by a feeling of happiness. It was wonderful seeing my friends after a year so far away. They had celebrated with me, we had partied together. At the same time I felt somehow detached from my home country, my family and my friends. During my long sojourn abroad I had missed parties, dinners and opportunities to offer real support when people I love needed it. However, I collected lifelong memories and had experiences I will always treasure. And straight away I am leaving again, to the next distant stop. And I want to. I want to see the Highlands, the Shetlands and the Orkneys. I want to get there. And I feel proud and privileged to be studying at such a globally renowned university.

Just before noon the plane lands in Edinburgh. I look out of the tiny window while we're taxiing to the terminal building. It's a small airport, nothing compared with Schiphol or Paris. I stuff the half-full pack of Marlboro in the puke bag in the little net in front of me, to be left behind. The stewardess announces that the doors of the aircraft are now open: *Here I go, entering a new phase of my life.*

I make my way to the apartment I found two weeks ago; a nice place on the second floor of a typical Edinburgh tenement on Gorgie Road, though this is not a salubrious address. It is a busy street in a working class area. The mixture of people suits me well but, more pertinently, so does the cheap rent in this very pricey city.

It had taken me a while to find a place. Daniel joined me in Edinburgh a week after the interview to look for a flat. We based ourselves at the Mortonhall Caravan and Camping Park on the south side of the city. We hadn't wanted to camp but there was a profound lack of alternatives – hotels and B&Bs were either booked out or ridiculously expensive. It had been August and the annual Fringe Festival was in full flow. It is the biggest arts festival in the world so the city was mobbed and there were wonderful street performances going on all over the town centre. It was enchanting.

However, Edinburgh doesn't need the colourful layer of artsy characters to mesmerise. It is simply one of the most beautiful, dramatic cities on Earth; a higgledy-piggledy warren of old streets and historic buildings, grand and quaint, together with massive slabs of greenery – including some genuinely quite wild stuff – makes up the Old Town in the centre. It is situated beside the elegant, expensive New Town – in which the "newer" buildings are mid-Victorian. The sense of the ages and the hilly vistas wow every student, tourist and worker who chance upon this beguiling city. Further, it is propped up and preserved by old money; Edinburgh town centre is prohibitively expensive for many people which – naturally and unfairly – is rather sexy.

The first flat we visited had brown wooden walls with purple and yellow furniture. "That's very nice," Daniel had said.

"Do you think I look like a bunny and want to live in Easter colours all year round?"

The second flat was painted deep blue and had a fishing net with shells suspended from the roof.

"But that one was nice, right?" Daniel asked, after closing the front door.

"I am not a six year old, and I am not planning to run a beach café. I hope to be a doctor, that place was ridiculous."

The next flat was doable, but had a sleeping bag and a piece of cardboard lying at the doorstep. The potential flatmate, who seemed weird, confirmed that most evenings a junkie sleeps there.

Then there was a small room in a "girl's house." The old nun welcomed us at the door by announcing instantly that I was not allowed to bring guys home, unless we were married. *Right… uh… hm…*

We visited a few more flats, which were either oppressively miserable or afflicted with other glaring problems. On the third day we finally found one to my taste but it was unfurnished and, apart from clothes and a laptop, I didn't possess anything. I phoned my parents and asked if they were willing to pay for some IKEA furniture on top of the bike they promised to buy me?

"I either have to step over a junkie every time I want to enter the flat and live with a freak collecting coke cans, or I need to buy my own furniture," I explained.

"Is there really nothing else on offer?" my mum asked.

"Yes, there is. But the landlady looks like Marilyn Manson dressed up as a nun, Jesus will be hanging in my room and there are Bibles lying around everywhere."

"Please go for the non-furnished option," mum replied.

My parents are not prone to pull out the cheque book at every chance but when we put on the screws they always come good. And I knew they would never countenance a Christian house. They had raised me as an atheist and always made it abundantly clear that religion wasn't their thing. When I was sixteen the new, apparently more readable translation of the Bible came out and I had asked to get it for my birthday, just out of interest. I wasn't planning on joining a church. I just wanted to have a clue what was written in the book that such a large part of the population finds meaningful. My parents told me if I really wanted it, I had to buy it myself. I went to the bookstore and noticed the Koran was significantly shorter and would leave less of a mark in my wallet, so I bought and read that instead.

"Does that mean you donate a few more euros to my existence?" I asked.

"This time, yes."

My mum had hesitated before she spoke. Daniel and I jumped on the bus to IKEA straight away to get some basics. We bought a bed, a dining table with two benches, a desk, a comfy chair, a few plates, cups and cutlery. By evening the apartment looked habitable though only the bed was assembled while everything else languished in those forbidding IKEA boxes.

I walk to the large window in the living room and look out over an educative farm for children. During the next three years I will hear the cocks crowing in the morning, the pigs chewing on metal chains and the cows bellowing. Occasionally I will stroll round the farm, when I crave fresh air and desire the company of dumb animals. However, I will mainly enjoy their presence from behind the glass window on the other side of the road. With one leg resting on the window sill and the other on the ground, I often stare in the direction of the farm. The overweight children on the premises display a remarkable lack of interest in the animals, at least before they have been turned into Big Macs, bangers and chicken nuggets. A few parents saunter among the enclosures on sunny days while the children gaze longingly at Snickers bars and Irn Bru in the vending machine. Even the impressive playground attracts very little interest from the kids; physical activity simply doesn't square with their shapes.

Next to the City Farm there is a pub with a sign indicating that all its patrons are Hearts supporters. Hearts, the oldest football club in Scotland's capital city, plays in Tynecastle Stadium, one hundred metres up the road. During the season they play, on average, every second week. On match days I like to lean out the window, watching the teenagers, families and elderly, all making the pilgrimage to the stadium. Very occasionally young guys burst into a short supportive song but it peters out quickly; generally the march is surprisingly quiet, faintly reverent. When the ball enters the net I hear an explosion of joy, but I can never tell which team scored. The primordial euphoria is presumably balanced by deep sadness from the opposing supporters, but I only hear the happiness. It will not be until the last year of my PhD that I will sit in the stadium, just once, watching an emotionally charged game between Hearts and Glasgow Rangers. That will be the day I learn many new Scottish insults, such as fuck trumpet, paedophile pal, shit gibbon, bloviating flesh bag and so eloquently on. Interestingly, traditional homophobic swearing is broadly unacceptable on the terraces nowadays, but new targets are under strident attack.

Five minutes after the match ends, Gorgie Road is overrun by armies of supporters who, despite having just attended a sporting contest, are a rather dismal advert for physical health. They invade the street such that taxis and buses struggle to get along it. Police are everywhere but actual incidents of note are rare. I look at the army of faces and try to figure out if its team won or lost. But nobody ever looks terribly happy or sad; just gratified in some ritual way beyond my comprehension. Half an hour later, Gorgie Road is back to being a lacklustre thoroughfare.

A few people stay behind in the Hearts pub. These are supporters who will keep me awake, but they will do so entertainingly. After a few hours in the pub, they mutate from football friends into uncontrolled, lurching meat-bags of testosterone. Some of them, while standing on the pavement having a cigarette, will lose the ability to stand, and make their way to Earth with tragi-comic stumbles. Others will shout and fight, glasses fall in the street, mobile phones fly, tempers rage and violent intent abounds. Yet it is still all rather unfrightening and, again, strikingly ritual; especially from my aloof vantage point. Amazingly few injuries ever result from the noisy shows.

The climax of the show – for me anyway, and I am about the most consistent member of the audience – occurs when men emerge from the pub with pool cues. The plan is to chase and batter each other, but running is a tricky business after seven hours of non-stop drinking. The tattooed guys might really want to smack each other's brains out, but carrying a weapon, let alone wielding it effectively, is an ambition beyond their talents after midnight. They fall over the cue, bang into the bus stop with it and – a personal favourite – they get the weapon stuck in the fence of the City Farm. They are reasonably adept at breaking the cue in half but that momentary act of concentration is usually as much as they can muster, and it seems to drain them. Afterwards, the fights fizzle out with one party leaving the scene, though invariably pausing to bellow a few sexual swear words from a safe distance. Homophobia re-enters the acceptable scope of insults at this stage.

Tonight, my first night after arrival, I am observing the first of these typical Gorgie Road scenes. It scares me and the thought that my flat is not ideally situated comes home to roost. But, hey, I have a home…

Chapter 4

It is seven in the morning. I have slept in my new apartment for two nights now and feel much less like a stranger. My eyes are wide open. I had an alarm set quite early as I know it takes time to get ready, but I haven't needed it. A healthy dose of adrenaline and excitement has wakened me for my first day as a postgraduate student. It would all kick off with a "School of Chemistry introduction event" for this year's selected students. Introduction days are not my forte. These are days that I have to be sponta-neous and smile, and not come across as a sociopath. I enjoy meeting new people but as soon as it is compulsory to "make new friends" I flop.

Nevertheless, I am motivated to go. I want to be a doctor. I want to be the first person in my family to have a title in front of their name. I fantasise about discovering a wonder drug which targets cystic fibrosis and writing an intellectually dazzling thesis about it, which will be discussed with great reverence amid prestigious cloisters and kick open doors to the most hal-lowed ivory towers in the scientific world. I can imagine myself walking down a red carpet to collect the Nobel Prize. Sometimes I even ponder how I will handle the inevitable fame… but I know my natural humility will see me through. Already, in the few days before I moved to Edinburgh, I started reading literature from the field and came up with some ideas for my ground-breaking research. Once the introduction day is done… Yes, I have itchy hands and want to get down to serious business.

Standing in front of the commode in the bedroom, I wonder what does one wear on one's first day in a chemistry department? I had seen most of the members of the McLean group during my interview and know they all

© Springer International Publishing AG 2017
K. Bodewits, *You Must Be* Very *Intelligent*,
DOI 10.1007/978-3-319-59321-0_4

looked okay – like Mark, not extremely fashionable, but also not as if they had plundered their grandparents' wardrobe. But then, the McLean group is working on the interface between medical biology and chemistry, making them probably the most "non-chemistry" group in the department. Today, I will mingle with the other groups; I am entering the world of chemistry geeks. It would probably be fine to show up like a hippie who has not had a shower for four weeks, or in a gangsta hoody, or in a T-shirt expressing offensive or sexist language – the bandwidth in academia is not limited – but a blazer would not go down well. I quickly stare at the only pair of high heels I have. I feel a love-hate relationship with them; aesthetically they are adorable, but they hurt. Anyway, high heels are definitely the ultimate no-no – too sexist or too sexy or too something or other. And they are practically ill-suited to a lab.

I look in the mirror that is still largely covered with packaging material and let my just-too-small turquoise miniskirt casually drop to the floor, after deciding it would be very unwise on the first day. I slide my freshly shaved legs into tight jeans and don a printed tank-top. *That's better!* I glide my fingers gently over the covers of my two new shiny notebooks before putting them in my bag – ready to go. I feel excited and nervous. Am I really clever enough to become a doctor?

As a teenager, I dreamt about going to university, sitting in a big lecture hall with my smart clothes and listening to a professor with small glasses resting on the tip of his nose. All the students would be very attentive of the bright man holding forth about difficult subjects hardly anyone can understand. We would all scramble after each other with interesting facts and ideas which the professor had raised. We would meet up in the student bar, drink a glass of cognac and discuss the latest news before going to bed. By the time I got to Edinburgh I had long ago realised I watched too many romantic movies about Oxford and Cambridge during my formative years. In reality we did not all spend our breaks talking about world affairs and not all students sat with rapt attention in the lecture halls.

However, there were some students who talked knowledgably about the world and culture, and who were even fascinated by the subject they were studying. They could talk about bird breeding, molecular interactions or the biological breakdown of pollutants for hours. And there were students who maybe did not want to pursue a career in science after their studies but who had a clear strategic plan of how to break into management consultancy, a traineeship at the community of Amsterdam or work for Heineken. The professors who taught me biology in Groningen might not have had glasses resting on the tips of their noses but they still lived up to some teenage hopes. They always came across as very critical, clever and skilful human beings. I often relished their lectures. They motivated me to study for my exams. Late in my studies, when I was working in their labs for months writing my master's thesis, or was remotely supervised during my times in industry, they infected me with their passion for discovery and they encouraged me to stay in science. Two professors in the Netherlands asked me to work for them but I wanted more, I wanted a famous university and the adventure of travel. By now, I had got addicted to moving. Addicted to meeting new people, learning new languages and finding my way around a new town. It makes me feel free, being anonymous somehow.

And so here I am, standing face to face with a statue of Joseph Black, a big cheese Scottish physician of the 18th century. While staring at the mouldering face covered with traces of moss, I think "well done dude, you made it!"

For the second time in my life, I walk through the long old corridor of the School of Chemistry. I pass the Museum again and note that this is where I must be in half an hour for the introduction programme. I make a left turn in the corridor, pass the chemical stores and up to the second floor. If I remember correctly from the interview a few weeks ago, I have to turn left after the stairs to reach the lab where I will be working. When I see an old door with a tiny opaque window, and adorned with a biohazard sign and the number 262, I know I am at the right place.

I open the door to see two people working at the bench in the middle of the same messy lab I saw during my first visit. It is a big enough workspace for four people. I greet them and wait while they slowly lift their heads and reply to my greeting with a cursory "Hello." Their expressions very clearly state, "What the fuck do you want with me?" *Ah well, it takes time to warm up with each other, maybe a teambuilding event or something…*

I walk towards the office and feel my backpack touching something. At race pace I turn to catch the bucket of ice I almost smacked onto the floor. Thank God there does not seem to be any samples in there. I feel slightly embarrassed by my clumsiness and keep my eyes on the ground. I take a step forward and feel my head hitting metal. *Ouch!* That was the huge autoclave I noticed on my interview day, when I had wondered in which century it might have been forged. That's going to leave a mighty bruise on my forehead. Feeling dizzy I step into the adjoining office.

There are four people sitting in the office, one of whom is Hanna. She is chatting with two other girls; their shabby, brown padded office chairs turned to each other. One is staring at a computer screen which is far from flat. In fact, this striking relic has much in common with the first computer my parents bought, in the 1980s. It had one game on it, which I loved, called Jumping Frog. I can't remember when they binned it but it must have been before we got the Internet in 1996. *Maybe they didn't bin it? Maybe they sold it to the University of Edinburgh?*

I say "Hi!" in my brightest, friendliest voice. Hanna and one girl reply, neither in a tone that suggests my brightest friendliest voice is doing the business today. The very small office is equipped with seven desks, three computers and one cupboard. The stench of stale cigarette smoke languishing in clothes – which I know too well – tells me at least one of them is a smoker. I look around to hang my coat and drop my bag, but all desks and chairs seem to be in use. "Which desk may I use?" I ask, brightly of course.

One girl, who has bubblegum pink hair cut in an asymmetrical bob that makes her face look much longer than it actually is, moves her head in my direction, opens her eyes as wide as possible, sniffs and says, "You don't have a desk."

A bit surprised, I enquire, "Is there another office where I do have a desk?"

"No."

I can't pin down her expression: does she feel superior or piteous? Either way this does not feel like promising rapport and my brain files her under "unsympathetic." I smile at her in a far from natural way. The girl who did not reply to my greeting and still has her back turned to me, says: "Maybe I will be gone in a couple of weeks and you can have my desk. But now I need it."

"Okay, great!"

I put my hand on the doorpost, sort of steadying myself and pausing to think: Is this real? Did I accept a PhD position without a desk? Was it even possible that the University of Edinburgh would *not* give you a desk?

"Can I maybe hang my coat somewhere for now?"

The girl turns her head in my direction for the first time. She has beautiful brown hair but it does not equate with her face. Her eyes are tinged with red and they lie deep in her face. Her skin has a grey-yellowish hue and her teeth are far from white. Her belly is hanging over the top of her jeans and she holds a bottle of diet coke. "You can hang it over mine if you want."

There is a black coat and scarf hanging over her chair, both covered with long brown hairs. When I come closer I realise the smell of old smoke, which fills the office, is emanating from these two items of clothing. I hang my coat over Diet-Coke-Girl's coat and drop my bag in the corner of the office.

"Is Mark in his office?" I ask.

I want to say hello, show that I have arrived, and ask him about the desk situation. I *must* have a desk, somewhere…

"Nope, Karin, he's at a conference. Good start for you. Holiday class two!"

This is from a short guy who has just entered the tiny office. He looks remarkably sloppy in a very wide jumper and talks with a strong Italian accent. Like Hanna, I'd seen him before during my presentation but I have no idea what his name is. As we certainly have been introduced, I think it's embarrassing to ask a second time and decide to wait until I hear someone else use it.

"What's class one?" I ask.

"You being on holidays yourself," he says giving me a playful wink. "But no worries, class one is a rather rare event."

"I've got thirty holiday days per year written into my PhD contract," I say, more to myself than anyone else.

"Yes, but you will be feeling too guilty to take them," Bubblegum-Bobline-Girl sneers.

I sit on a chair hoping that one of them might initiate small talk with me. None do. For all they heed my presence I might as well be in Amersfoort, or Timbuktu for that matter. Feeling utterly superfluous to requirements, I listen into their conversation. And I feel the right side of my forehead swelling; the pain of the bang is still reverberating. I took time to pick sensible clothes this morning, I wanted to fit in, but I am marked as the clumsy girl with the signature of an autoclave on her head.

Feeling like a five year old who has had her favourite balloon burst by laughing bullies, I walk to the Museum a couple of minutes before the programme starts. About another twenty-five PhD students stream in around the same time. I'm delighted to note they look much less nerdy than I expected. In fact there are two guys in the room that are really quite cute. We all shake hands but again it's all too quick to remember names. A dark-haired guy, who doesn't seem much older than anyone else in the room, stands in front of us and starts instructing everyone about the games we have to play. He is at least thirty pounds overweight, attired in jogging pants and a white T-shirt that looks like it has been starched and ironed by his mummy. From the expression on his face, he clearly feels very important standing there. There is nothing appealing about this guy. As soon as he starts talking, with a strong Glaswegian accent, I not only struggle to understand him, but I also struggle to contain the irrational hatred he has inspired in me. "Aye. Ye jist standing, tryin to find the person with the same wee card…"

I want to stuff the stack of cards he holds in his left hand into his mouth to keep him from saying another word. Extremely demotivated I take the card out of his hand not really knowing what to do with it. Somehow this guy got so far into my allergies that I only focused on him, the irritant, and

I entirely neglected his instructions. I ask the girl standing next to me what the plan is with the cards, and after her explanation I move to the right spot in the room. About an hour into the intro games, I pretend to go to the bathroom but don't return. I had found myself in a line-up with twenty-five other new PhD students, ordered by body length. We also did a line-up based on our hometown's distance from Edinburgh. Fascinating as height and distance surely are, I felt I just didn't fit in and exited lest anyone rumble me as a party-pooper.

"They said what?" my mum asks, sipping from a cup of coffee on the other side of the screen.

She had asked me to phone her after my arrival to update her on how I was getting on. But until this afternoon I didn't have internet in my apartment, so I had only sent a quick I-am-alive text message before.

"It turns out I don't have a desk or a computer," I say into my laptop for the second time, trying not to sound too frustrated.

For a few seconds my parents look dumbfounded on the screen. I'm not sure if Skype is just delayed or if they are thinking about an answer.

"Can't you bring your own laptop? Then at least you have a computer."

"I could, but I can't do anything with it. I went to the IT Department and they don't want to connect my private computer to the network for security reasons. Unfortunately, there's no Wi-Fi in our lab."

My mum is clearly thinking now; she's a problem solver, always has been.

"This sounds too strange. Of course there will be a desk and a computer for you. They really won't take more PhD students than there are desks. Everyone needs a desk nowadays. Trust me. "

"It didn't sound like it, mum. They all told me I won't have a desk."

"Just wait until this boss of yours is back, doctor McLean I mean…"

"I hope so. I wouldn't know how to do my PhD without it."

"Really Ka, don't worry about it."

"I hope you're right, mum. I really hope you're right."

After Skyping with my mum, I am convinced that the misery I felt during the first day of my PhD will give way to a whole new situation once Mark is back, and somehow I am excited to see how it all unfolds.

Chapter 5

The second day wasn't much different from the first. I spent both days perched on one of the lab chairs resting my back against the rusty autoclave. I had printed a new stack of papers to read, but I struggled to concentrate. No one talked to me unless I asked them a specific question. It felt miserable, and I ended up clock-watching. I didn't have anything to do and hadn't even talked to Mark about where to start.

Today would be different; Mark will be back from his conference, and I am ready to roll. Full of positive energy, I jump onto my bike before eight and ride to campus. I drop off my coat and bag in the corner of the small office, as I have done on previous days, and walk straight to Mark's office down the corridor. Through the slit of the door I see that his office is dark. I knock and there is no reply.

"He doesn't come in early, he's usually last of all," Hanna says when she sees me staring at the clock.

"But he will come in, right?"

"Sweetheart, just enjoy the time he isn't here!" says the Italian guy, who I now know is called Erico.

At 9:30 a.m. the heavy wooden door of our lab opens.

"Hi Karin," Marks says, good-humouredly. "I see you settled in already!"

"Yes," I say happily, assuming he is being ironic. But I remain a tad uneasy.

"Good, let's get you started!"

He pops his head into the office that is completely filled with people. "Hi, I'm back."

© Springer International Publishing AG 2017
K. Bodewits, *You Must Be* Very *Intelligent*,
DOI 10.1007/978-3-319-59321-0_5

No one looks too excited. "We missed you," Erico jokes, puncturing the tension in the air.

"I bet you did."

Erico and Mark briefly discuss the latest football results, like two friends meeting up for a drink after work. I have always felt envious that guys have this universal topic for small talk. It sounds dull as ditch-water to me but it is unquestionably a fail-safe. I have watched thickos and geniuses delight in each other just because twenty-two men kicked a ball about a slab of grass some days previously.

"Hanna, you've got a student starting tomorrow on the fatty acid project" Mark says after the football topic fizzles out.

"Me?" Hanna asks in disbelief.

"Yes, you," Marks says, impatiently ticking his key chain on the doorpost. "Have you ordered the primers for the project yet?"

"No, I didn't know I would get a student to supervise."

"Get it organised, so we don't waste time," Mark sighs deeply and shakes his head in disbelief. "And you should show Karin around the hospital." Hanna nods.

"Have these guys shown you the lab yet?" he asks, looking in my direction.

"A bit, yes," I lie.

"All of you, get her started!" he says looking round the office but, alas, no one takes it personally.

Mark walks in the direction of the door, indicating that I should follow. We enter his office and he empties a chair for me. He's full of positive energy as he had been during my job interview. He talks a lot, and has this disconcerting habit of laughing when there is nothing to actually laugh about. He is a bouncing bunny enthusiastically showing me a plethora of scientific papers while expounding research ideas. Having read myself into the field a bit, I am at least able to partly comprehend what he says. Then he speaks about the other lab members, and does not hold back on criticism – unsparing criticism. I have no reason to doubt what he says. To me he is the only normal person in Lab 262, the only one who is actually talking to me and one of the few who does not faintly disdain my mere existence. Encouragingly, he adds several times that for me there is nothing to worry about, I will be different... different from the others. "You will do better! We will conduct great research together!"

The projects he wants me to start working on are quite different from what was written in the original PhD description, which was also discussed during my interview, and which he received funding for. But, thankfully, the big picture is unchanged; I will still be working on a super bug killing people with cystic fibrosis – just the proteins I will be researching are completely different. I will work on a different biosynthesis pathway than originally planned but I don't really care. After listening to him for over two hours, and feeling exhausted by all the information, we order the first set of DNA primers together. I can get started in a few days when they arrive.

Before leaving his office I gently enquire if, per chance, I happen to have a desk somewhere? And if I might, just possibly, be granted access to a computer? "A desk you can share. You don't need to sit in the lab, sit in the office. Like the others."

"Eh... eh... okay... And a computer?..."

"You don't need a computer."

"I don't need a computer?" I meekly respond, trying not to sound as if I am questioning his sanity.

"You should read the papers from the lab," is his only elaboration.

Am I dreaming? My mind is racing. *How could I have forgotten to ask during my interview where I would be sitting, and what the computer situation would be? But then why would I have asked? During my job interview and in the original project description, it had been clear that the PhD would include a lot of bioinformatics – that is what I had been doing before, as a master's student, and I want to expand on that. We did talk about that, right? I had a virus making my body dysfunctional to a certain extent but I'm not suffering from Alzheimer's. I recall my interview! That said, I didn't write a protocol of the interview. I received the offer of penetration from Quasimodo's twin brother during my stay but I did not receive a contract with a detailed project description... Mark did most of the talking during the interview and now, reflecting on it, he mostly waved the few questions I asked away with rather non-specific remarks like "yes, no worries... you get all that..." Maybe I was clear enough that day that I wasn't interested in shrivelled genitals yet failed to express how important bioinformatics is for me? Maybe I wasn't clear enough in expressing my wish that the project exists in reality?...*

I clutch at the doorway, anticipating the worst. I even sense bile rising in my throat and fear my face is losing its colour. My belly starts to cramp but I nod passively at Mark, turn around and walk back to the lab. I am feeling tricked by empty promises.

For hours I sit with my back against my rusty fossil friend, just gazing around the lab. I resolve to be positive. I reflect that had I coded for Stephanie Shirley in the early 1960s I would have gotten by with pencil, paper and telephone. That consoles me not one jot. And I go back to hours of muttered expletives and fantasising about asking Mark "what the hell century are you living in?!" At five on the dot I unlock my bike and cycle to my apartment. I call Daniel to complain.

„ No desk, no computer?
What the hell century are you living in?! "

"Good you are calling. I've got fantastic news, but I didn't want to phone you so early; I thought you wanted to check out this swimming pool after work today."

"Yes, I did. But I am too tired from sitting around the whole day doing nothing. I can do it tomorrow night."

"What's up?"

"Turns out that indeed I do not have a desk or a computer. And the other PhD students... I'm not sure. Mark says they are all hopeless. It is all kind of strange here... Whatever, I shouldn't be complaining, I'm lucky to be here and have the chance of a PhD at this university… How has your day been? What's the news?"

"I will come to Edinburgh!"

"Really?"

"Yes. This professor I told you about in the biology department, he finally replied… And I am welcome to come and do an internship."

"You're kidding."

"No, I'm not. And I phoned the University of Groningen straight away and they are happy to give me the credit points for it."

"Wow, you've been active," I say, knowing that it would normally take Daniel quite a few days – or weeks – to pick up the telephone and sort such things.

"You think I would want to be separated from my doctor-to-be for three years?"

I could imagine that… In contrast to me, Daniel lacks the ambition to become a doctor. He follows me wherever I go like a loyal puppy, which is touching but also indicative of the fact that he has nothing much else to do with his time.

"Have you got a starting date yet?"

"I'm free to pick, meaning that I will come soon. I looked at flights and I can be there as early as Sunday afternoon."

"Cool!" I say so excited that it surprises me.

After spending an intense year in Shanghai together, part of me was happy to have some distance, especially as our relationship had been getting more serious by the day and I strongly doubt we are a good match. But another part of me is looking forward to having him here – just to have someone familiar around. *Oh the passion.*

"We should celebrate!"

"Definitely. I'll buy a bottle of wine for Sunday evening and cook some-thing nice."

"Love you."

"Love you too."

Chapter 6

As the lab stool is back-breaking, I start to play musical chairs, hopping from one temporarily free seat to the next. I sort of read paper after paper, which is to say I flick through a few PhD theses written by my predecessors; mainly perusing the acknowledgments page and looking curiously at the number of publications in the appendices. Some don't contain any, others only a few. As publications are the key measurement in research performance, this is bewildering, faintly alarming. Oh well, I convince myself, I will indeed be different, as Mark had said.

I feel lonely, have nothing to do really, and start to regret that I weaselled out of the introduction day at the first opportunity. I wish I had been more patient and played along, because then I might know someone. As it is, my breaks are no less lonely than my "working" hours.

To occasionally escape the work I am not doing, I stroll around the department. In contrast to our lab, most labs have windows facing the corridor. This allows me to look inside and observe researchers, mostly at benches with fume hoods. Some labs are well-equipped and very modern, while others are kitted out with *olde worlde* apparatus like our gothic autoclave. The rigidity of academic hierarchies, which would make Medieval Japan look like a hippie commune, is writ large in the size of the lab and the equipment therein. The status of the person running the lab is as plain as a peasant smock or a gilded robe.

On the ground floor there are at least two gilded robes. I had never heard their names before but they have super-cool status stuff, which they may or may not need but I bet they like having it anyway. They also have lots of serfs working for them in their tidy, airy, spacious labs. I will soon learn that

© Springer International Publishing AG 2017
K. Bodewits, *You Must Be* Very *Intelligent*,
DOI 10.1007/978-3-319-59321-0_6

they are professors who were actively courted to boost the university's reputation. Apparently they make an impact, at least in the scientific community.
That will be me one day, I muse, while I dodge the autoclave en route to
my non-existent computer.

I am clearly not working for a gilded robe. I'm beginning to wonder if
he's even a peasant smock. Perhaps he's threadbare underpants. His enthusiasm, youthfulness and size of his research group appealed to me. I didn't
want to sink into insignificance or jostle for the supervisor's attention in a
large group. Now I am not sure if this was smart thinking, but at least Mark
is eager to get things going.

Passing the small chemical stores, I remember Mark telling me this is the
place to get myself a lab coat. It has an aura and familiarity which is comforting; shelves filled with glassware, pipette tips, tubes and solvents. There
is the inevitable small room filled with containers for dry ingredients. In the
back left corner I dig through the lab coats wrapped in cellophane and find
one my size. Despite the straight cut of those uniforms making all human
bodies look equally squared and yet saggy, like a granny's bathrobe, this
uniform will prove my work is real and serious.

At the counter a good-looking young guy sits behind a computer filling
in order forms. Another, older grey-haired man has a friendly chat with the
lady in front of me. I understand what he says but, as far as my ears are concerned, every syllable coming out of her mouth is gibberish. She has red
curly hair pulled tight by a hairband, a black pullover and a very round ass
stuffed into jeans that are about two inches too short. The white sport socks
atop her black leather shoes slightly disturb me. *Chemists.* She issues forth
some syllables, laughs raucously and waltzes off, whistling. This is a happy
shop, a world away from the lab. The older man greets me with such a
friendly smile that my loneliness evaporates. He scans the barcode on my
lab coat.

"To which grant shall I charge it?"

"I don't know."

"Do you have a stores card?"

"No," I say with polite, pleading eyes, which make no difference at all.

He gives me another charming smile, though I think this one is tinged
with exasperation. My loneliness is reforming as he explains how the system
works. I need to get a card with a grant number so they know where to
deduct the payment.

"Okay, I will ask someone in the lab and come back in a sec."

Up in the lab I finally learn something. All of a sudden, my new colleagues are willing to talk to me. It is carefully explained to me that on the

ladder of respect our lab is subterranean. And my miserly stipend does not extend to a lab coat, or any money for equipment or other consumables for my research – my tuition fee and my "ample" monthly salary of £1100 is all that it covers. It transpires that only two out of the eight students have any access to grant money. And neither wishes to splash out on my lab coat. I had to literally beg for a card to take to the store. What I don't know yet, not at this stage: today is just the start of the begging, and in future I will have to beg and beg and beg, just to buy basic stuff for research. I appear to be at the bottom of the pecking order in a sub-respect lab run by threadbare underpants man.

"That took long," the older man says.

"Apparently my grant doesn't cover anything like this, so I had to ask a few people how it works."

"Yeah, not all grants cover consumables."

"I would have expected that all PhD students are the same."

"Nope. Totally depends on who is funding you. Even the salaries differ from one student to the next."

"I didn't know that."

"You'll learn," he says, nodding with compassion.

"Bring this as well, you need it; then you will not need to ask for a card again so soon." He places an A4 sized blue lab book on top of the lab coat on the counter.

He gives me the same friendly smile as he did earlier, but this time it barely affects me. I feel helpless and small. I take the stairs back up to the lab. When I see the autoclave, I want to climb in and die. Instead, I make my way downhill from the King's Buildings to a small newsagent and buy a ten-pack of cigarettes. I walk a bit further down the road to Blackford Pond, which I presume is man-made. I plump onto one of the wooden benches and watch the ducks. I try to console myself with the veritable fact that the only way is up...

On my return I find myself desperately hoping that my colleagues do not smell my consoling Marlboros. Part of me knows this is a long shot because, in my absence, they have not all had their noses chopped off. Still, the futile, self-deluding desire to go undetected is an essential part of the smoker's social survival kit.

I start browsing another thesis but cannot concentrate at all, not even to find the acknowledgements page to sneer at. The room is ten square metres. It's too full with eight PhD students and now, for no clear reason, several undergrads are busying about too. I stare out of the window, watching people depositing empty chemical bottles in a container. The antique autoclave is hissing away and the cacophony of voices never lets up. The two most mature PhD students, Bubblegum-Bobline and Diet-Coke-Girl, continuously whine about how they hate this place and haven't been paid this month. After three years, their stipends ran out and Mark does not believe they have enough results to write up their theses. Oh well, at least I am not the only person feeling vexed and defeated…

In the evening I sit on the window sill, my laptop connected to my speakers. I listen to the song Hey There Delilah by Plain White T's, over and over again. Once I can no longer bear to hear that plaintive yearning I call up Queen with I Want to Break Free. Scientific breakthroughs and dreams of fame seem far away, on Planet Ridiculous.

Chapter 7

"Would it be an option to walk a bit faster?" I say to Daniel, who is walking half a step behind me in the direction of the university campus.

Autumn has arrived suddenly, with bitterly cold mornings. It has been exactly fifteen days since Daniel came to Scotland. Today would be his first day at campus. He won't be greeted by the gothic face of Joseph Black in the morning but instead will enter the modern, swish Darwin Building, to accrue the final credits for his master's degree.

It is a few miles from my Gorgie Road flat to King's Buildings, but as Daniel didn't get round to buying a bike yet – *he only had two free weeks in Edinburgh* – we opted for walking together.

"Don't be so hectic, enjoy the beauty of the city, the smells of autumn coming…" he answers dreamily.

Such comments often make me wonder how Daniel came to study natural science. He would be a much better fit in a social science or arts faculty, where talking about feelings and emotions is an important part of the daily business.

"It's Monday morning and I have actually got work to do, Daniel."

I started my research project as soon as the stuff I ordered for my research arrived. Thus far it is yielding no results, so, Mark gives me tasks so I get familiar with the lab equipment. From sitting around passively like a chicken expunging an egg during my first week, my PhD quickly transformed into a challenging full-time job. Apart from Bubblegum-Bobline-Girl, the other PhD students started to thaw towards me. At least that was my impression, but maybe their form of communication – or rather the paucity of it – is slowly becoming my new norm.

© Springer International Publishing AG 2017
K. Bodewits, *You Must Be Very Intelligent*,
DOI 10.1007/978-3-319-59321-0_7

Mark talks to me, a lot. Invariably when I am tired and want to go home he comes into the lab and buttonholes me for hours. I see the other PhD students leaving, grabbing their coats as soon as he appears at the door, but I'm staying. He talks quickly and loudly and constantly. His monologues are intense, piercing, self-dramatising, but at the same time deeply enthusiastic. I try hard to follow everything he talks about but sometimes I don't even know if he is talking about my project or someone else's. It is difficult to contribute and I am not sure if I am even supposed to. His conversational rhythms are so unrelenting and bereft of pauses. The flow is all one-way, forceful, unstoppable and dark like a river delta without the allure. Sometimes I seem to be watching and listening from a distance. At other times I feel like a sounding board, just an object soaking up this torrential flow of words. Most nights, I leave with cramps in my stomach, a bad headache and the sinking feeling that I've been violated. But I convince myself I am newly charged with research ideas. Mark sees a rosy future for me – my research will be groundbreaking!

"I'm not walking slowly, Ka," Daniel states.

"You are," I say, turning towards him and closing the chest clip of his backpack.

Daniel looks surprised. "Now, move your arms to support your steps, and we will go double the speed," I add.

Daniel laughs it off and starts to move his arms as I instructed him in an exaggerated fashion; we do indeed go faster. When we arrive at the chemistry department we say goodbye. Daniel wants to give me a kiss before heading to the other side of campus, but I decline. I hate being kissed in front of university or anywhere really when the setting is new and the kiss has nothing to do with love; just the mere male need to publicly mark territory. "Let's keep it professional," I say, pushing him away from me.

"Have a nice day," I add.

"Right."

Daniel has delayed me a fair bit, so it is almost 9:15 a.m. when I finally arrive in the lab. I drop off my bag and coat, get into my lab coat and walk straight into the cold room, which would easily fit two bunk beds and a small desk and is always kept at four degrees. There are no windows, and the smell inside is awful. There are shelves filled with buffers and chemicals that need to be kept cold, plates with bugs and several machines running almost round the clock. Like the rest of the lab, it's a mess. I take the rack holding the one falcon tube I had left here yesterday. There are white flakes swimming through a colourless buffer inside, like out-of-date milk curdling into coffee. *Shit, that doesn't look good.*

"Should I check this with Mark?" I ask.

"Mark? When does he do experiments in the lab?" Hanna asks.

She regards me like I have seriously messed up. "I have no idea… I just started."

"Believe me: Mark is not the right person to ask."

I look surprised.

"He's in his office, he hasn't been working in the lab for years," she adds.

"The imprint of his ass is engraved in his office chair," Bubblegum-Bobline-Girl's voice shouts from the office.

She must have overheard our conversation. As our small office is directly attached to the lab, you always hear what's going on and being said everywhere.

"He doesn't have any skin on his ass anymore from sitting too much," she adds.

"Oh now it's getting interesting! How do you know what his ass looks like?" says Quinn, a six foot Englishman with short, light blond hair, in the third year of his PhD.

Bubblegum-Bobline-Girl replies, "Don't be so literal! I don't actually know if he suffers from decubitus! But I suspect he does."

"You don't need to defend yourself, but let's talk about it over lunch."

The guy, and several other people in the office, are grinning. Under other circumstances, I would have put my head round the office door and joined in this rare moment of levity. Instead, a worried smile is spreading over my face. Hanna holds the 50 ml falcon tube in the air, and looks at it a second time.

"Did he really say you should add so much salt to the buffer?" she asks.

I check the loose piece of paper lying on top of my brand new lab book. With my finger I move over the sloppily written notes while reading them out loud. I stop in the middle and turn the paper in her direction.

"It really says 900 mM NaCl right?"

She takes the sheet of paper out of my hands. She looks at the scribbles for a few seconds.

"Could be, but it could say 300 mM just as well. Which would be just about right."

She looks at me with her large brown eyes. I look down at my slightly too large lab coat. It still shows the wrinkles of coming just out of the box. *From this piece of new clothing everyone can tell that I am a newbie, obviously a keen newbie, but still not someone you need to take seriously.*

"Is there any way to desalinate it?" I ask, in despair.

It has taken me a full week, including the weekend, to purify this protein that I don't need for my research. Mark told me it would be good to purify it anyway, to familiarise myself with this particular technique which I might need later on. I loathe pointless exercises. I am a scientist. I look for meaning and reason. I find the mere idea of having to spend more days and late evenings in the lab redoing this futile experiment quite upsetting.

"No," says she. "It has all precipitated."

She points at the white flakes swimming in the tube. She hands me the tube and says: "I'm afraid you might have to redo it."

"But… it will take me days, and I don't even need it for my research."

"Then don't do it!"

"What can I do? Mark told me I should do it."

"You should only consider doing something after the third time he has told you to do it," chips in the female voice from the office. "Otherwise you go mental here."

"And only consider it; considering doesn't mean *doing*," Erico adds in his charming accent.

Hanna nods in agreement and walks off to continue her own work.

A great lesson, I have learned something useful here.

I look at the tube one last time, in the way you might look at an unwanted gift from an ex-lover. This tube contains the only proper chemistry work I have done since I arrived. I let it drop into the yellow waste container at my feet.

For a moment I am staring at the chemical waste containers outside, below our lab, unsure what to do next. My plan for the next three days has been screwed by the precipitated protein in the bin. "If you have time you can join me going to the hospital today," Hanna says. "I've got stuff to do there; I could show you around the cystic fibrosis lab."

"Thank you," I hear myself say. "When will we leave?"

"Now."

We put on our coats and walk to the bus stop close to campus. "You'll like it in the hospital," Hanna says.

"You did your master's there as well, right?"

"My bachelor's. I don't have a master's degree."

"You don't have a master's degree?"

"Nope."

"How did you get this PhD position?"

"In the UK you don't need to have a master's to enrol in a PhD programme; a bachelor's is enough," Hanna says proudly.

"Really?!"

And we all end up with the same title?! It's just that some skip two years of study in the process, apparently. WTF…

"Schools are short in Scotland too. You can finish high school and start studying when you're only seventeen. So if you don't take breaks, you can be finished with a PhD as young as twenty three. Like Stacey, the Scottish girl next door in the Baxter group; she's only twenty one and in her second year."

"Maybe the chemicals in Irn Bru bring out your inner Einstein..." I mumble bitterly.

"Oh no, that just makes people fat."

Within ten minutes we arrive at a glistening white building just outside of town, surrounded by agricultural fields. The autumn sun reflected by the building hurts my eyes as we walk to the main entrance – it looks dazzlingly *Brave New World*. We enter a large open hallway with a high roof. In front of us, on the first floor, there is a large library with bookshelves reaching to the next floor.

"Right goes to the hospital, left to the labs," says Hanna, taking a left.

We go up the stairs and arrive at the library but we walk in the opposite direction. She takes a card out of her wallet and swipes it through a card reader next to a glass door. "You need a card to enter; otherwise patients could stroll in."

We walk through a long corridor with a carpeted floor and into the last office on the left. It is large, hosting at least twenty people. "This is my desk. If you come to the hospital, you can drop your stuff at my desk. As you can see there are no places free here either."

Of course not. In this hidden slave world, a lowdown newbie cannot dare demand a desk, good grief no... So you have two desks and I have none... I nod.

We drop off our stuff and go to the labs on the other side of the corridor. We walk through a lab full of impressive machines and then through another glass door. "This lab has biosafety level II, meaning you can grow pathogens which are moderately hazardous to humans. It is important here that you always keep the door closed; we don't want the bugs to leave the lab. And this is not Lab 262; you really *do* need to wear a lab coat and gloves here."

Hanna hands me one of the lab coats at the door and points at a box with purple gloves next to the sink. There are large benches on one side of the lab, and intriguing equipment on the other. There are a few people working, but most benches are empty. It is tidy, organised and new, sort of heavenly to my jealous eyes. "Quite different from the mess in the chemistry building, isn't it?" Hanna says, upon seeing me gawp.

"It is indeed."

"It's a nice place to escape to."

"Yes. I believe you."

"Those three benches belong to James' lab. This is my bench. We will have to share it, so please keep it tidy."

She rests her hand on the one in the middle and then indicates the next bench: "That is Brian's bench, the postdoc Mark talks about a lot." *Oh yes, I've heard his name so much I am bored of him before even meeting him.*

"He must be really good," I say.

"Right," Hanna says, clearly unconvinced. "Anyway, don't use his bench. He will go mental if you do."

There is a girl laughing at the last of the three benches. "You really don't want to touch stuff on top of this bench either," she says.

"Have you met each other?" Hanna asks.

"No, we haven't."

At this point two girls get rid of their gloves and jump off their lab chairs to reach out their hands to me. One of them has long reddish hair bound together in a ponytail. She is tall and skinny, and her hands feel as soft as silk bed sheets. "Leonie," she says, smiling.

The other one is almost a head shorter and has shoulder-length, thin blond hair hanging sadly over the collar of her lab coat. She has a large blue bruise on top of her eyebrow and fresh scratches on her chin. Her right wrist is covered in a bandage but nevertheless she uses her right hand to shake mine. Her skin feels much rougher than Leonie's. As I say my name I notice a dried-up drop of blood leaked from her ear. "Sharon," she says as if she herself is not a hundred percent sure of her name. She must have noticed that I'm observing her with caution and smiles reservedly, almost shy. *What kind of arsehole is this girl with?!*

"Karin is the new victim of Mark's relentless monologues in the evening," Hanna says, looking from Leonie and Sharon to me. *New? Victim? Hmm…*

"Oh, poor you!" says Leonie.

"It's not that bad, just his timing is sometimes a bit awkward," I say. "Are you working for James?"

"Yes, I'm a postdoc in his lab," Leonie says happily. "And sometimes I'm also his personal secretary… like copying stuff for him…"

She says this playfully, rather than resentfully, but Sharon adds wryly, "It's because you're female. Brian is the apple of his eye!…"

Sharon explains that she is a technician working for James, mainly analysing saliva samples from patients, "Checking if they caught any infections and checking the antibiotic resistance of their newly acquired bugs, and stuff."

She points at the piles of Petri dishes standing next to Leonie. I tune out trying not to stare at her bruises – so incongruous with her demeanour and friendly disposition. I am relieved when Hanna indicates the tour is continuing. She opens a drawer and takes a key out. "Come, I'll show you where we keep the bugs."

I follow her and as soon as we are far enough away from the glass door, I ask, "What's wrong with Sharon? Did you see? Is…"

"You mean the bruises?"

"Yes!…"

"Plays rugby, always injured."

"Oh thank goodness! I thought she got beaten up."

"Well, she did, that's rugby."

We go down two floors to the basement. Hanna carefully instructs me that the key always has to be returned to its rightful place; the university keeps a lot of bugs which are potentially dangerous to mankind, so we don't want evil loonies from the outside world coming in with anarchic ideas. We enter a basement room that reminds me of The Svalbard Global Seed Vault in Spitzbergen. It hosts eight large white freezers, each with a display on the door indicating a temperature of minus 80 degrees. "A404 is ours, so no need to touch the rest. And you *need* to wear gloves, no bare hands here."

Hanna sticks the key in the keyhole of the freezer and turns around. A blast of deathly cold hits my face. "We have to hurry. The alarm goes off when the temperature rises above minus 68."

There are about fifteen drawers, all fully covered with mysterious flaky icicles which make the letters on each hard to read. "We have got the lower two drawers, N and O."

Hanna pulls on one of them. There are at least five hundred different tubes carefully organised inside. All tubes have a number and Hanna tells me which ones to take. When lifting the tubes up they immediately freeze to the gloves, and I need to use a second hand to prise them off and place them in a special rack Hanna brought along. By the time we have gathered all the samples, which takes us less than two minutes, and closed the heavy door, the display indicates minus 71 degrees. "What we have here are the mother stocks of all bugs you might need for your research. Only use them once, and create your own stocks, so you don't contaminate those."

Walking upstairs, Hanna points out where to get buffers, freshly poured plates, bottles and anything else I might need. She shows me which media to use for which bug, how to isolate DNA and she gets me started creating my own "superbug stocks." We happily chat with Leonie and Sharon, who seem curious enough to get to know me. They tell me about lab conflicts, unwritten rules and competition between researchers. It isn't all positive, but I do get the feeling that I'm doing something useful, that I am learning stuff and I am surrounded by people who treat me like a human. I sense it as a divine reprieve from eternal disappointment. I want to stay here.

Mid-afternoon we troop back to King's Buildings campus. When we enter the office of Lab 262 Mark is talking to one of the new project students.

"Hey, what's up?" he asks as soon as he sees us.

"Well, the protein I purified precipitated overnight…"

"Where is it?" he asks, shaking his head from left to right.

"I binned it."

"You shouldn't have done that, you should have shown me!"

"Sorry. Hanna showed me around the hospital instead, which was very nice," I say, desperately trying to steer the conversation onto something agreeable.

Mark nods at Hanna. "Good. But you don't need to spend much time there. I want you to do your research here."

"Why?" I ask surprised, thinking about the nice people, large benches, fancy equipment, clean lab and general glory of the Brave New World.

"Because I am here, and I can help you," states Mark, and promptly makes to leave.

"Mark, wait a sec," Hanna says, tentatively drawing his attention. "I just wanted to tell you that I booked a holiday with my family in February for a week."

Her voice is almost trembling. Mark sighs and ticks his keys against the doorpost. "It's up to you, it's your PhD," he snarls, and walks out.

I sit down on one of the temporarily free chairs, and notice that I am out of breath for no reason. I am trying to persuade my lungs to work at a normal speed while my mind plays ping-pong. *Is he serious?*

It is early evening and the weather has changed from stunning sunshine to pouring rain. Daniel has been waiting for me in front of the chemistry building until I am finally ready to go home. I'm at least forty-five minutes later than agreed, because Mark came in to tell me about a new research idea he had, just as I was packing up. The sharp teeth of the tone he demonstrated this afternoon was largely gone, and he spoke with enthusiasm about an article he had read – something I could follow up on. I listened to it with the lesson I'd learned this morning in mind; *only consider working on a project when he has mentioned it at least three times or you will go mental here.*

I left the lab the moment Mark captured a straggling project student. *You really have to run with the pack at the end of the day or the predator will get you, and bore a hole in your soul…*

"Sorry I'm late," I say to a soaked Daniel.

"I phoned you three times! Why did you not pick up?"

"Because Mark was talking to me, so I couldn't."

"Couldn't you tell him that I am waiting here in the pouring rain for almost an hour, and you just need to tell me that you're late?"

"Honestly, I'm not sure if I could."

"What bullshit! Who are you working for? Stalin's long lost son?"

"How was your day?" I ask, to change the subject.

Daniel's mood improves instantly. We make up on the way home, while he tells me about the PhD students, his supervisor and the lab itself. He's got a desk and Wi-Fi access. He speaks with much more enthusiasm than I did after my first day. His PhD colleagues don't seem to work too hard; they spend much of their time playing Facebook games and chatting, which doesn't seem to bother Daniel in the slightest. By the time we arrive home, he is slowly changing topics. He is talking, I am listening, passively, as if I am not there. He tells about an uncle who has just announced that he intends to leave his wife and two children in order to live with a younger women round the corner... It should interest me, but my mind is drifting off... I leave the room, get into my sports shoes and run through town, like a distraught little girl who wants to flee...

Chapter 8

"Hello."

"Hi," I reply.

A strange pause occurs. It may be rhetorical and is definitely confusing.

It's Lucy, the Wallonian girl from the lab standing next to me while I sit on a stool holding a pipette through purple gloves. I carefully place the tip between two glass plates while making sure I press the blue stained protein solution in the right slot. After pulling the tip back she is still standing there, a tad too close methinks, holding nervously on to the bench. Lucy is an outsider, like me – never involved in lab conversations; for hours at a time it's as if we do not exist. I'm not sure I have ever heard her speak except when answering very occasional questions from Mark. I turn my head in her direction and lift my eyebrows.

"I actually went to the Pear Tree Bar yesterday, it was very nice," she says in a firm voice.

Having never properly communicated with each other, this seems like a peculiar gambit, for strangers anyway, but hey... chemists. I look at her and am quite fixated by her large green eyes and her perfectly symmetrical jawbones. For the first time I notice that Lucy is astonishingly beautiful – a beautiful weirdo.

"Really, who did you go with?"

"With a few people from the Johnson group."

"Okay."

"I thought maybe you might want to join us next time?"

Fucking finally! Someone to go to the pub with! I don't care how weird you are... I don't care if men will look straight through me and even push me out of

© Springer International Publishing AG 2017
K. Bodewits, *You Must Be* Very *Intelligent*,
DOI 10.1007/978-3-319-59321-0_8

the way just to catch a glimpse of you… I don't care if you're a Neo-Nazi sex perv who kills babies for kicks… ANYONE is preferable to the life of a scientist with nothing but a lazy boyfriend evening after evening…

"Why not? When are you going?"

"We might go for a few drinks to KB House after work today."

Why on Earth would you want to go to that dreary dirt-hole on campus after work?!

"Great! I'd love to join you. Thanks."

"Okay, I'll let you know when we leave."

I manage to nod casually while my heart leaps for joy. The blissful day I've been waiting for, fantasising about for weeks, has finally arrived. I got invited to join people for drinks, just when I had concluded I was as welcome in the world as Mussolini at a reggae festival. One evening not in the lab… not in my apartment… not strolling aimlessly through town…

I am almost three months into my PhD and I am doing – mostly – meaningful work. Like all other first-year students I gave my introduction presentation at the School of Chemistry lecture series. Both Mark and James had been happy with it. Hanna had shown me some techniques and took me to the hospital a few times. I like her but she seems socially saturated with long-lasting friendships in Edinburgh. And even if she wasn't, I'm not sure our personalities are suited to hanging together outside of work. I got introduced to Brian, the postdoctoral golden boy, who I subsequently encountered several times and pegged for a frustrated dictator. Existing as he does in the obvious and repellent bubble of research failure, he gets ever so wound up about everything I touch or use. Information he supplies is – magically – bereft of at least one essential puzzle piece, and this slows my progress. He is clearly afraid of being overthrown by Hanna, Leonie or me, which is paranoid nonsense since we are not even working on remotely similar projects. Maybe he had a toy taken away from him once too often in childhood and is still in trauma. Who knows? Who cares?…

Despite Brian providing a rough edge, the hours spent at the hospital feel like mini holidays compared to the School of Chemistry. The extraordinary tensions in Lab 262 befit high drama, yet they are generated by the petty matters of money and space – or, rather, the lack of both. The constant struggle to get access to equipment and consumables is enough to make a strong woman weep and a good woman murder. We have several shopping lists hanging in the lab, filled with stuff we need but every time anything is added, it is followed by a lively discussion about the allegedly consumerist behaviour of certain lab members drying up lab funds. It's all balderdash and flummery; nobody is terribly wasteful. Our lab is simply dirt poor.

The financial stress is compounded by a nasty element of interpersonal power-plays owing to there being just two people with access to grant money. The big shots downstairs conduct ground-breaking research with fancy new kit while we waste precious time and life trying to get by with out-dated, inefficient, old hand-me-downs. Mark just sent Erico and Quinn down to northern England to fetch a second-hand table top centrifuge that Marie Curie would have called an antique. It occupies a large part of the lab but it works, up to a certain speed, which is all we dare hope for. The very basic equipment triggers a Boy Scout instinct, which I somehow like, but at times it feels like we're stuck behind the east side of the Iron Curtain smelling the bananas of the big shots a floor down.

Bubblegum-Bobline and Diet-Coke-Girl spend less and less time in the lab. Some weeks they call in for merely a day. For me it feels like two chickens less in a battery cage, but it frustrates Mark. "They don't have enough results to become doctors yet!" he keeps on telling me.

In the evenings, the jingling noise of Mark's key chain betrays his entrance and resounds like a gun for the race to start; all the other PhD students grab their coats and race off, with practiced stealth. This we-all-happen-to-leave-at-the-same-time ritual seems too obvious to me, for now. Within a few months I will be a devout follower.

With shocking examples, Mark shares his concerns about the competency of the other lab members with me, over and over again. At first I feel sorry for him, having landed so many rotten recruits from the vast ranks of applicants. Due to his stories, my colleagues – the people I am learning from – are not blank slates anymore. I see them as ropy, workshy second-raters giving me suspect guidance. At the same time, I hear my colleagues complain about Mark with no holds barred. They laugh behind his back about the advices he gives. They warn me that the projects he has in mind for me are not possible within the timeframe of one PhD and the equipment we possess. And I fear they are right about this. He expects me to clone, purify and design novel assays for nine different types of proteins of a complete biosynthetic pathway – even for my non-expert ears this sounds like over-reaching on a megalomaniacal scale.

Independently, Bubblegum-Bobline and Diet-Coke-Girl convey markedly similar tales of Mark not reading thesis chapters that have been on his desk for months. *Oh didn't he? Really? Of course, he has got a lot going on... and from what I have heard about your work... I suppose, if anything really took his eye, then he would...*

They both talk about swerving round Mark to Prof. Gilton, Mark's old PhD supervisor, and asking him to read their theses instead – just in order to leave this lab astern forever. I feel shunted between the two sides and have no idea who to believe. Perhaps deep down I know who the rotten apple is, and I am in denial.

I am committed to my research, so most of my waking days I am at the Chemistry Department. I leave the house before Daniel gets up and often do not return till just before bedtime. The majority of the time in the lab I am working but sometimes I am just avoiding going home. Daniel sitting at my IKEA kitchen table each evening, quietly smoking my last cigarette, is not spicing up my life. I dare say I'm not spicing up his either. Some spice is needed somewhere... His bags are still unpacked and lying in the small corridor of our flat. Every day he takes out the pieces of clothing he needs, no more. It has transformed the almost empty flat; the beautiful wooden floor has become a sort of playground where you have to jump from one island to the next. I've got good. But I am bored of the game and I am hoping Daniel gets bored of it too. I would be equally delighted if he just decided that airing his clothes by way of a sojourn on the floor was insufficient in cleanliness terms, and that he really needs to start washing them. But it's early days; one of these happy events could yet occur.

Regardless of Daniel being there or not, I feel lonely, but after tonight this could change. I will have some social contact, just like a real grown-up with a real life.

With a smile on my face I start to move agar plates to the flame. After the second plate I quickly walk to the "list of consumables to buy" hanging on the door and write "Ampicillin" adding an exclamation mark in the vain hope of conveying some urgency. The list is almost full, with restriction enzymes, chemicals, pipettes and other everyday stuff. *That will lead to another discussion about who finished it. I don't care, it's all...* Suddenly I hear yelling in the office. It doesn't take me long to realise that it is Bubblegum-Bobline-Girl and Mark having yet another argument, but today it's all-out nuclear thunder, even by their own high standards. I had not heard either of them enter because of the radio and the background noise of machines running. Bubblegum-Bobline-Girl is not the type of person to mince her words, not to us or Mark. The verbal violence between her and Mark is a show to behold, captivating in a train crash way.

"Your thesis is worthless without it!" Mark shouts.

"Thanks for your input, Mark! It's an idiotic plan and I'm not going to do it!"

I know this will be the end of the argument. The note of finale chimes with the familiar assertion.

I take a pipette from the rack and, like the other PhD students, pretend to work so it isn't too obvious that we have all been listening – as if we had a choice. Mark storms out of the office with a red face, boiling from anger for sure, but perhaps also from embarrassment. He stops just after passing me. He turns around and fixates me with his bulging eyes that so far, when he talks to me, expressed energy. Today they express pure rage. He barks "Why is the water filter not being exchanged, Karin?"

My heart lurches. *Where is this coming from?*

"Um. What water filter?" *I am such an idiot, we only have one. This will make him furious?*

"What water filter? What water filter?! How long have you been here, Karin?"

"Three months." *He's not looking for the literal answer, stupid girl.*

"And where do you get your water from for your experiments?"

He looks at me as if I am some form of life akin to the slime on a snail's belly.

"From there," I say, pointing at the water purification system.

"The light on the machine has been red for weeks, indicating that the filter needs to be exchanged. Why did you not do it?" *I've never seen the light green during my time here.*

"Because I didn't know I had to." *I am really digging my own grave here.*

I feel everyone is watching me. For the first time I see my colleagues, from the corner of my eyes, expressing sympathy for me.

"You didn't know you had to? Does someone need to spell everything out to you?!"

Mark is getting even more red-faced by the second as he leans towards me. I inhale deeply and feel my face getting red as well.

"No, but..." *I really should stop digging.*

"It's my fault, I'm on it," says Logan, my tall Northern Irish colleague, while closing in on me and Mark. "We don't have filters anymore. I'll get the order in today."

Logan started his PhD just a couple of weeks after me. We haven't talked much. Logan is more of the quiet type and, like Hanna, did his degree in Edinburgh. He seems nice enough as a colleague but he is well dug in at the Uni and in Edinburgh, socially saturated, like Hanna. I am pretty sure he has never seen the light being green on the machine either. He's just trying to help me, out of decency.

"Right!" Mark says, nods to Logan and storms out.

"Thanks," I whisper, after the door falls closed.

"No worries."

I feel my heart pounding in my head and sit down on a lab chair trying to take long breaths. The entire tirade had taken less than a minute, probably even less than thirty seconds but it feels like I've just run to the top of Arthur's Seat. *What was that all about? Why did he marginalise me in front of all those people? What on Earth did I do wrong?*

I feel small, downright belittled, a tricked and trapped fool now bound to play the fawning, gawping subject at the court of King Dullard.

"Welcome to Lab 262," Quinn states, resting his eyes on my face, pressing his lips together.

During the lunch break I quietly escape to the changing rooms of KB House. I change my jeans for shorts, slide my feet into my sport shoes and run away… I run away from university, away from my fellow lab-mates, away from the maddening hierarchy of King Dullard and his PhD jesters…. I don't fit in, it's my fault, I feel this. Tears well up in my eyes, but as long as I keep my eyes wide open they are dried by the wind before they can run down my cheeks.

It is almost 9:00 p.m. when I open the door to KB House. I order myself a pint of Guinness, plop down on the seat Lucy has kept free for me and slump down.

"He talked to you for ages," says beautiful Lucy.

"He did."

It had been two hours ago that I quickly signalled to Lucy that I will join them later. Like most evenings, Mark came in to talk to me. Though the atmosphere was different from other nights, Mark pretended that nothing had happened in the morning. For me it was weird, but at the same time reassuring that he still trusts me. He talked with enthusiasm about all the projects he has in mind, complimented my work and gossiped about the other students. I tried to listen, but as it was already well past seven I couldn't decide whether I was just exhausted or ravenously hungry. He went on and on and I just kept on hoping that he had other plans for his Friday night, something other than just talking at me. But he didn't seem to have any.

"Don't worry, he will lose interest in you," Lucy says.

"Did he use to talk to you every evening as well?"

"No, he has never been really interested in my project. And I started at the same time as Babette; her project is Mark's baby, so he talked to her."

"I notice he doesn't talk to Logan either."

"Nope, Logan is just a space-filler."

I lean forward to catch a glimpse of the handsome guy ordering at the bar.

"He's good-looking, isn't he," Lucy whispers.

"Gorgeous. Have you dated him?"

"Yes. He's got a girlfriend though…"

"Nice one."

"She doesn't live in Edinburgh."

"If it's a different post code then it doesn't count…"

Lucy laughs. "True. He's weird anyway. I went to the cinema with him last week. At the end of the evening, instead of a moment of intimacy he opted for crossing the city in the middle of the night to a particular 24/7 in order to purchase one cucumber, one tomato and one carrot to make up for the lack of vegetable matter in the food he had consumed that day."

"Wow!"

"I bet he sent a message to his mum afterwards to tell her he'd been a good boy."

"Is he good in bed at least?"

"Oh, I don't know. I left him with his carrots."

"I also wouldn't want to have a rabbit in my bed."

We giggle like old friends.

"You've got a boyfriend, right?"

"I have."

"What is he doing?"

"Dangling in the purgatory of eternal studentdom," I mumble softly.

"Right…"

"That was mean. He's doing a research project for his masters at the Biology Department."

"Will he do his PhD after?"

"No, he doesn't want to."

"Will he stay in Edinburgh?"

"I guess he would like to, but the British economy is in tatters and good employment is a long shot for an indolent non-native speaker without any relevant experience in anything…"

"Haha, you really like him, don't you?"

"I used to, but I'm no longer sure."

"Do you smoke?"

"I do, but don't have any on me."

"Have one of mine, you must need it."

We walk outside. I feel tipsy from downing a full pint on an empty stomach and decide I should definitely get something to eat after the smoke. Lucy points at the thick book sticking out of my backpack.

"What's that?" she asks with a mischievous smile.

"What do you think it is?" I reply half smiling.

"Thomas Bright's thesis."

"You read it as well?"

"Mark wanted me to."

"Did you do it?"

"I flicked through."

"I really don't want to read it. But he keeps bringing it up. Pretty much every evening he asks if I've read it. But it has nothing to do with my project, so I don't know why."

"It's an experience as such, just go for it. Because I seriously doubt Mark has ever read it himself!"

I feel grateful for the nicotine and support. Two unfamiliar guys join us outside, fellow smokers. "Wow, two pretty girls waiting for us outside. Life could be worse…" says the one with curly blonde hair, and a nice Australian accent.

"Your blue scarf really fits your eyes," the same guy says to me with a cheesy voice.

"And your yellow shirt really fits your teeth," I reply, wondering who I am insulting.

As the bar is on the university campus, chances are that it is someone working here.

"Oh, oh, ooh," says he, grinning in a way that doesn't suggest he is really offended. "I am Harold. Nice to meet you."

"Are you working here?" I ask.

"Nope. I'm just visiting this dude," he says holding the shoulders of the other brown-haired, average-looking guy.

"Yeah," says brown-haired dude. "Harold is a senior lecturer in Glasgow. We know each other from way back. We're collaborating on a research project."

"What are you girls doing? PhDs? Postdocs?" Harold asks.

There's something about this guy. He's not so terribly handsome, but he's got something about him. Maybe it is just the good-humoured smile.

"PhDs," Lucy and I reply at the same time.

"What field?"

"Biochemistry."

"Which group?"

"The McLean group."

"Hahaha. Poor you!"

"Why?"

"His reputation leaves much to be desired. Everyone knows him, even in Glasgow, but not in a positive way!"

"What's bad about him?" I ask.

"Two buzzwords: unreliable, incompetent. *But*, I have never worked with him so I don't know what's true about it and what isn't."

Shit.

"You girls fancy going to town?" he adds.

I quickly look at Lucy who seems slightly less curious about this pair than me. "You want to go?" I ask her.

"Why not? If you want to…." she says with a noticeable lack of interest.

"Okay, we'll come."

"Great! That's probably our cab right now," says the average-looking guy indicating a taxi pulling up.

Lucy and I quickly fetch our coats and join the guys in the cab. "Any preference as to where we go, ladies?" Harold asks.

"I don't mind," I say. "So long as it's not a club."

"Whistlebinkies?"

"Okay."

We drive to the underground pub where a band is playing very loud Indian pop music. We order a pint and chat about the whole PhD experience thus far. It doesn't take too long before I am feeling too tired to be sociable or funny. I tell Lucy that I haven't eaten yet and really need to go home. She orders a cab.

Before going to bed I sit at my desk in the living room eating a pack of chocolate chip cookies Daniel bought. Curiously and dutifully I am reading some of the thesis. It is of little interest and absolutely no relevance whatsoever. However, there is a noteworthy gulf between what the thesis says and what Mark says when he is trumpeting it. *Wow, he actually hasn't read the thesis he bangs on about at every opportunity, how bizarre…*

Before bed I quickly check my email. One email from Hanna with the subject "Christmas dinner" asking if we all happen to have time on the 7th of December.

"Seventh works for me," I reply to the email, fully cognisant that my availability will not be pivotal in the final decision of the Christmas dinner date. I have long understood that, as a first-year PhD student, I am the appointed runt in the poorly contrived wolf pack of Lab 262.

I close the laptop and make my way to bed. In the faint street light through the curtains I look at the silhouette of Daniel's sleeping face on the pillow next to me. I quietly whisper, "Just cheat on me, darling. That would make it all so much easier."

Chapter 9

Duct tape. I need duct tape! Holding the bands of my hold-up stockings between index fingers and thumbs in order that they do not roll down to my ankles, I play "island jumping" to the living room. I press my legs together to free my hands and start digging through the toolbox in the large lighted cupboard. It is not there. *Of course it is not there… nothing is ever where it should be…*

I issue a few random swear words and feel Daniel's eyes on my back.

I hear a distinctly outdoor voice, and realise it is mine: "Where's the fucking duct tape?!"

"Jesus, relax."

"There are only so many hours of relax time available in one house and you've taken all of them!"

I now realise that I really do need assistance. With gargantuan effort, I continue, "Sorry, I'm just late. Do you happen to know where the duct tape is, by any chance?"

I wanted to sound conciliatory, but sarcasm triumphed.

"It's there," he says, pointing to the upper shelf in the cupboard.

"Can you do me a favour and get the scissors?"

Daniel raises himself from the office chair, so bloody s-l-o-w-l-y and ambles to the kitchen. I hear him opening a drawer, then another one.

"It is hanging on the magnet bar," I shout, struggling and failing not to sound annoyed.

He returns with the scissors.

"And now?"

"Now we are going to tape these tights to my legs. What else?"

© Springer International Publishing AG 2017　　　　**63**
K. Bodewits, *You Must Be* Very *Intelligent*,
DOI 10.1007/978-3-319-59321-0_9

He shrugs. "Of course… I should have guessed. What else would you do with duct tape? How do you want do it?"

"Once around and three times vertical."

I hold the tights up and indicate where he needs to stick the tape. He rolls the tape once around my upper leg, half on the stocking, half on my skin.

"Not too tight!"

He cuts it off. Carefully he takes three short pieces and sticks them on vertically, over what we just taped. He starts the same procedure on my other leg while enquiring, "Is it normal to tape stockings to your legs with duct tape?"

"Of course not! But I don't know what else to do. I'm a scientist, not a lingerie expert."

"But you bought them."

"Yes, I bought them," I say with a glare to ward off more irksome comments.

"Why did you not buy normal ones? Like those that go over your arse."

"Just let it be, Daniel!"

He laughs mockingly and points at my duct-taped legs. "Very sexy."

I pull my dress down to just above my knees, narrowly covering the tape. I get my only pair of high-heels from the bedroom and, with some apprehension, slip into this alien footwear. An uncomfortable tingle shoots up my legs.

With some difficulty I totter on to the packed bus and find a seat near the back. The moment I deposit myself in it I feel, and even hear, the tape releasing from my skin, on both sides. The stockings slither down me with remarkable ease. *Oh shit, did anyone see that? What will I do?* I slip out of my stilt-shoes and, at very high speed, pull the stockings over my toes, crumple them together in my fist and deposit them in my handbag.

The bus drives along beautiful, spacious Princes Street overlooked by the lighted Nelson Monument on Calton Hill. It turns right over North Bridge where history surrounds you in the different layers and characters of all the old buildings – a dramatic jumble of the ages, and yet always those huge patches of greenery in the city centre; deathly black at night. *Oh how I love this town in the dark. How lucky I am to live here.*

When I disembark the cold wind cuts through my skin. It would have been grim with tights but this is downright tortuous. I ring the bell at 17 West Nicolson Street. Lucy buzzes me in from three floors up and I ascend the ancient wooden stairs.

"Hi, how are you?" she says, looking drop-dead gorgeous in a simple, classic black skirt with crème-coloured blouse and thin black tights.

Out of breath I reply: "Fine. Cold legs!"

"Oh, you don't have tights?"

I take the tatty tights with duct tape out of my hand bag and show them to her.

"They were supposed to be hold-ups but they did not stick."

She pulls the tights apart and regards the duct tape. "Those are not hold-ups, Ka. Those, you wear with a garter belt."

"Garter belt?"

She walks to her laptop and types "garter belt" in the search box. The screen fills with pictures of elegant women in beautiful lingerie. "Okay, yes, that is quite sexy. Quite different from how I feel."

It is deflating. *Are scientists just not sexy? Was Marie Curie sexy? Wait, Lucy is sexy. She's the freak…*

"Wine?"

"Sure"

We drink and smoke with our upper bodies hanging out of the two tiny windows next to each other. Michael Jackson's childish voice can be heard in the distance. *Sometimes the best part of an evening out is that period of hope and expectation between being ready and leaving.*

We chit chat about everything and nothing and decide to walk to the restaurant. It is just after 8:00 p.m. when Lucy pauses in front of a bright blue facade. The small windows and the large white lettering are homely and inviting. Our colleagues' faces are shining in sweet candlelight within the premises.

"Here we go," says Lucy.

I empathise as Lucy inhales deeply and then sighs before opening the door. Apprehension stalks the Lab 262 Christmas Dinner. It doesn't help that I am wearing a dress, stilt-shoes and no tights in the brutal Scottish winter. The moment we enter the small restaurant, a handsome, young waiter greets us with a welcoming smile. "We belong to that group," I state, pointing in the direction of my colleagues.

Everyone is tastefully attired in ironed shirts, smart trousers, stylish dresses and neat skirts; all rather chic yet suitably wintry, apart from the stupid Dutch girl with bare legs.

Mark greets us with a friendly smile. He looks pleased and excited; this is a Mark I haven't seen lately.

After a few minutes of Mark being unquestionably charming, my mind is in turmoil: *Maybe it is because the grant proposal was successful. Mark told us yesterday that he managed to get some new funds for the lab, and he did seem very happy about this achievement… Maybe this is why he has been so moody and unpleasant since the water-filter argument?… Maybe it's just the pressure of academia, a world where you completely rely on third party funding, negotiated by way of long dreary documents, endless competition and, above all, who you happen to be chummy with... Maybe Mark has just been having a difficult few weeks, getting stressed and depressed about funding... Maybe everything will get better now... or maybe I should steel myself against wishful thinking at a Christmas party...*

Mark looks around the table triumphantly and indicates to the bar that we are complete now. The waiter hands out menus and before he can turn away, Mark touches his arm and asks, "Drink?"

His eyes flit from me to Lucy and back.

"I'll take a red wine."

"Me too," Lucy says.

Everyone at the table is drinking wine or beer. The atmosphere is strange. I can't work it out. People talk in a friendly way while tension and excitement vie to dominate. On the one hand, the Christmas excitement evokes memories of childhood and celebrations of years past, as Christmas is wont to do; I think of me and my sisters, three happy little girls bouncing about at the breakfast table, unable to contain ourselves as we stare over at the presents under the Christmas tree, all bursting to find out what Santa brought us this year. But on the other hand, this evening also calls to mind one of those stifling dates with someone you definitely do not fancy and are beginning to dislike, before the starters have even arrived, and yet you feel it would be imprudent to just drink your way through this slow evening because alcohol is likely to render you rude and undignified. Ominously, I finish my glass of wine before dinner is served and Mark orders a few more bottles...

The waiter places a steak in front of me, and it is leaking blood like fresh roadkill. *Damn, I asked for well done!* I consider asking him to cook this dead flesh a bit more but I wimp out. I feel the red steak mixing with the foundation of wine in my stomach. Like a good chemist, I conclude that the liquid I have imbibed might not be the best solution for dissolving this bit of cow. Slowly my head starts to spin and soon I am afraid I will end up on the floor. I resolve to flush the flesh down with water. Mark orders bottles of booze the moment one empties. Everything starts to normalise as the evening progresses. At least it looks that way to me as the consumption of alcohol oils the engine of bonhomie and dampens cares about dignity. We are laughing and joking with each other in a way I have never once seen in our lab. It's nice, I'm starting to enjoy myself, that's confusing. We pool money together to pay the bill and for drinks thereafter. Mark insists on taking us to a pub in Leith.

"Why is that?" I ask Lucy while smoking a cigarette in front of the restaurant.

"He always wants to go there."

Standing next to us is Bubblegum-Bobline-Girl, displaying bland Celtic tattoos on her feet. Thus far she has never uttered one friendly word to me and, since she is soon to finish her PhD, I assumed she never would.

"Mark loves Leith," she says, and actually smiles at me. *The alcohol must have you in its power and I hope you are going to regret your friendliness tomorrow when you wake up; because I know I will definitely regret being friendly to you after all the unfounded "fall to death" looks you have thrown at me in the lab.*

"Why is that?"

"He lives there, and believes it is the best place on the planet. In fact I am surprised he agreed to come here for dinner. It's always been Leith in the past."

Vaguely I recall Mark telling me during my interview that if I was going to look for a flat in Edinburgh, Leith was the best place to live. I checked it on the map at the time and concluded it was too far out of the city centre and too far away from the university. The rest of the lab members join us outside. Like a swan guiding its uncertain cygnets, Mark leaves the restaurant last.

"We need three taxis," he states, and jumps onto the road like a rugby player ready to deck a prop forward. He very abruptly halts a passing cab. Four of us get in.

"Port O' Leith," Mark instructs the driver.

Wow, that sounds exotic...

We drive off, leaving the rest of the lab members behind. We cross South Bridge, then North Bridge and wait at the traffic lights before turning onto Leith Walk. On the little square in front of us, a guy in a kilt and full Scottish regalia is playing bagpipes. Like me, he doesn't wear any tights but he has thick, high socks which leave only his knees exposed to the cold. There is a hat in front of him to donate money and I drunkenly wonder if I could pay him for his socks...

We travel down Leith Walk, a famous street I haven't seen before. We pass a few pubs and clubs, and it gets dingier by the block, until there are only small, seedy shops and dark, residential buildings.

"We're there," announces the driver indicating a red pub with white metal bars in front of two large windows.

A few middle-aged drunks totter about at the entrance. From the car you can already hear that they are engrossed in a passionate conversation about politics and life, which would make no sense to them were they sober. As it is, they are all certain of their genius. They look surprised when we exit the taxi.

"Hi girls, welcome to the best pub in town!" one blurts, opening the door for us in faux gentlemen style.

There is an old lady standing on the other side of the door, obviously an habitué of the Port O' Leith, perhaps even the owner. She seems to be attired for a carnival party; pink velvet dress, boa and high lurid red shoes. Damaging lifestyle is writ large on every inch of her face. She takes my hand with her wrinkly fingers, each of which sports a ring, in a way my grandmother might, and presses her heavily lip-sticked and botoxed mouth to my cheek.

"Welcome," she says.

Though overwhelmed by this warm and yet frankly revolting largesse, I try to reciprocate whilst disengaging from her grip without cutting myself on one of those knuckle-duster rings. *I feel soiled, I hope I did not just get a weird disease... Then I will never be able to finish my PhD, I won't walk down the red carpet like a celebrity when I win the Nobel because I will be oxidising in a 24/7 care home...* I am momentarily appalled at my own snobbishness. But only momentarily...

„ What the hell are those girls doing here?
And how come they still have all their teeth? "

The pub looked small from the outside, and now it looks small from the inside. It is just a local alcoholic pub with an exotic name. The roof is covered in a mix of English, Scottish, sport and random pirate flags. There is a preposterously large selection of alcoholic drinks on the shelves behind the bar and most customers are at least fifteen years older than me – or least look like they are. Early graves beckon for these ghouls, yet in here, tonight, they have no cares. A few middle aged ladies, with coloured, greasy hair and dresses that reveal far too much wrinkly booby flesh, regard us with disbelief; as if to say, "What the hell are those girls doing here? And how come they still have all their teeth?" I share their bewilderment. *Why on Earth would Mark bring us to such a place?*

The others arrive and Mark orders beers all round from the money pot. I check the glass to see if it's dirty and take refuge in conversation with Lucy. Then Mark joins us. "You like it here?"

Just at that moment a very drunk middle aged man emerges from the gents with his zip open and saggy, limp member on casual display.

"I love it here," I reply.

Mark, now also looking at the penis, says: "It is a strange place. We go here every year. It's become a tradition."

As if he is too curious to wait and see who comes out of the gents next, he heads to the toilet. The music is loud. The second burst of alcohol really starts to kick in, and we find ourselves dancing between the tables. When Mark comes back he shouts at the bartender: "Those girls want to dance on the bar!"

Do we?! Really?!

The gaunt, grey-haired bartender smiles. "Just climb up!"

Mark turns to us and says: "It is the only pub in town where you can dance on the bar."

"You want to dance on the bar?" I ask Lucy.

"Sure."

That wasn't the answer I anticipated. But I can never resist a challenge. So we find ourselves clambering on to the bar by way of a chair. It's an unsteady, unnerving business when you're doused in alcohol and sporting a short dress. I get a ludicrous sense of achievement when I stand up on the bar even though it feels a bit wobbly at first. Something yields inside me – my self-respect perhaps – and I feel comfortable, as if dancing on a bar to entertain prime specimens of the Edinburgh underclass has always been an ambition. We get full support from our lab members, all cheering wildly when we start dancing, and playacting of course, to *You're the One that I Want* from *Grease*.

"Somehow I knew we needed the wine before," I say to Lucy, when the song ends.

She laugh, "Yes!"

Mark says, "If you keep on dancing you get your PhDs!"

I know he's just trying (and failing) to be funny in a boorish drunken way, but our inhibitions have also dissolved in alcohol and Bubblegum-Bobline-Girl, who joins us on the bar instantly, shouts in my ear, "If he would just read my fucking thesis I'd get my PhD! I am working for free; does he really think I should do some lesbian lap-dancing for him too?!"

I exchange knowing smirks with Lucy and we even manage a reproving smile at Mark.

Two more colleagues join us on the bar, and a couple of songs later every student member of Lab 262 is there; on the bar in this seedy hovel dancing like there is no tomorrow. Mark is the one lab member not on the bar, but in fact that's fair enough since he is already unsteady on the terra firma. During *I've Had the Time of my Life*, from *Dirty Dancing*, I propose Lucy tries out "the lift" and, quite suddenly, I know I really have had *too* much to drink now. Lucy talks me out of trying it but we dance and laugh plenty more. For the first time since my arrival in Scotland, I love Lab 262.

Chapter 10

Consciousness is upon me as soon as the alarm goes off. I'm not sure I was fully unconscious for much of the past six hours in bed. Even during supposed sleep, my experiments are on my mind. In my sleep, I had run through the streets of Edinburgh, to the university to save the samples that I forgot to put in the fridge the day before. When I had arrived at the university I realised I had forgotten my card, but a Batman-like figure appeared and helped me through a window. I feel physically tired and mentally confused. I'm happy to escape my quasi-dreams and get my day started.

I get out of bed and go to the bathroom. On my way I see that Daniel is already in the kitchen. It is unusual to see him up this early.

"What's up?" I ask, chewing on a toothbrush.

"You have been kicking around the whole night. I couldn't sleep."

For a few seconds I wonder if I should say sorry, which would have been nice, but I decide there is no reason to excuse myself for something I can't control. I nod instead, sympathetically I hope.

"I made you a fresh orange juice and the toast should be ready in a minute."

"What a luxury!"

I walk over to Daniel and give him a kiss, then to the window and look out over Gorgie Road. It is still pitch dark outside, and it is at least three hours till the sun rises, if it rises as such… It is the end of January, the middle of Edinburgh winter, and there are many grey, foggy days that see little light. Some of the Lab 262 inmates are laid low by the lack of light but it doesn't seem to have much effect on me. Daniel is coming towards me.

"You got a lot to do today?" he asks.

© Springer International Publishing AG 2017
K. Bodewits, *You Must Be* Very *Intelligent*,
DOI 10.1007/978-3-319-59321-0_10

"No, not just at the moment. I have to start my experiments from scratch because they didn't work."

"You want to come over for lunch today?" he says resting his hands onto my shoulders.

"I could do," I say, looking him in the eye.

There has been a lot of tension between us for over a month now, since Christmas really. I'm getting tired of it. I guess Daniel is too.

The Christmas holiday was raw stuff. I had known that it had incensed Daniel when I packed a crate full of stuff to read over Christmas. But it never occurred to me to care. His parents had given short notice that they had rented a cottage in the Highlands for a week to spend family time with Daniel and his sisters. I've never been a big fan of Christmas holidays but for the first time since I moved out, at the age of seventeen, I had looked forward to spending the holiday with my own parents. To go back to the village where there is the Christmas and New Year tradition of "carbide shooting," whereby the town alcoholics buy themselves explosives and use them in conjunction with milk jugs to create a deafening bang. It's horribly dangerous but somehow they never seem to sustain major injuries. We would hide from the heavy, flying lids inside the house and sit around the plastic Christmas tree. We would exchange presents which we had all hand-picked for ourselves. My family doesn't like surprises so all of us specify exactly what we want, what it costs and where you can buy it. This year I had looked forward to the predictability, the low excitement and being surrounded by reassuringly familiar characters. The absence of fuss does not deplete the ritual and love, for my slightly eccentric family anyway.

When I had announced I might visit my own family instead, Daniel had gasped at me as if I just confessed to child murder. His parents had paid a fortune at the last minute to rent a cottage large enough to afford private space for Daniel and me. They had rented a second car, presuming I would be joining them. Though I didn't want to join, I felt strangely relieved that I did not need to face my family: How could I honestly answer the questions they would ask about my PhD? My parents knew that I did not have a desk and no access to a computer, but they were unaware of the full grimness of the situation — and oblivious to my lack of research results. They were still on cloud nine about their daughter doing a PhD at a highly ranked university, doing groundbreaking research on cystic fibrosis. They were talking about it as if I was on a wonderful, noble mission which would save and enrich lives. I knew my dad would beam like a Cheshire cat whenever someone enquired what his daughters were doing. They probably did not have to enquire; he always contrived opportunities to brag about us as children. He is, like any good dad, embarrassingly proud of us. How could I tell him his youngest daughter, who just graduated with excellent marks, has so far not generated one remotely significant result nor learned anything scientifically useful. Tail between legs, I followed Daniel and his family to the Highlands.

For a week I had more or less locked myself in the cottage bedroom to read scientific papers, coming out for food and showers. A few times Daniel begged me to partake in family hikes, and I did. I avoided talking to his family as much as I could, and only exchanged necessary information. I could not stand being there. I could not stand them. When Daniel and I were alone in the car back to Edinburgh he shouted at me. I had misbehaved and disrespected his family. I had never seen him so angry. My initial reaction had been defensive: "I have work to do."

"Over Christmas holidays, Ka?!!"

"I don't have a clue what others do, but I want to be a successful researcher, Daniel," I had shouted back at him.

Then I had apologised, because I had behaved selfishly, no doubt. But at least the reading had given me plenty of new ideas. Feeling replenished, I began again on the second of January, with new methods … I tried… and tried… and tried… Nothing worked thus far. Mark was putting the squeeze on me. He was starting to expect meaningful results, and I had zip. The most noteworthy biochemical reactions in my world were those taking place inside me, owing to frustration and stress.

The stress of work never helped any relationship, I know. But when you are blinded by stress, it's hard to know where work problems begin and relationship troubles end; they blur and cloud the mind and the only certainty is, I discovered, that selfishness runs amok. Daniel doesn't deserve it, he is a good man, so today I oblige him and agree to go for lunch. It might help puncture the tension between us.

I spread some *E.coli* cells onto the agar plates, place them in the 37 degree stove below one of the benches and make my way to the Darwin building where Daniel's lab is on the eighth floor. It is spacious, much larger than ours. A stereo blares out Coldplay's piano ballad *The Scientist*, so Chris Martin's falsetto fills the room with whimsical longing. It's quite carefree, as befits the mess. This is different mess from the mess in Lab 262. There it is cheap, shabby mess. Here it is relaxed and rather expensive mess. Also in contrast, the benches in Daniel's lab are – gosh! – separated. This strikes me as a magnificent triumph of aesthetic design, though really it is just a bunch of workspaces with room to breathe and concentrate. On a shelf above one of the benches there is an A4 piece of paper informing everyone: "I will break your fingers if you touch that." Arrows point down to different spots on the bench.

"Wow, that is quite direct," I remark.

"Ha ha, yes it is," says a short blond-haired guy, lifting his shoulders.

"He is not British, is he?"

"No, he isn't."

"How do you know?" asks a girl with long brown hair.

She is also smiling. *My God, people are friendly and happy in their university lab!*

I reply: "Because it would say: 'I will break your fingers if you touch that. Terribly sorry, excuse me, I was only joking.'"

I walk towards Daniel on one of the last few benches. He is wearing gloves, pipetting a PCR product into an agarose gel.

"I still need a couple of minutes," he says.

"No problem."

I take a seat on one of the lab chairs and out of boredom open the lab book on the bench. Only the first two pages are filled. One is dated the start of October, just after his arrival. The other is dated today. *After spending eight hours per day over a course of four months in the department, Daniel has conducted enough experiments to fill precisely two pages in his lab book. I am with someone who takes four months to do two PCRs. [Polymerase chain reaction – basic stuff in the 21st century, requiring less than an hour of hands-on work.]*

"It seems that the probability of me catching you working in the lab is slightly less than me winning the Nobel. Yet in fact it happens – today. I guess I'm lucky."

"Hmmm…"

"In all fairness, I already noticed on Facebook that you do devote an impressive amount of your time in the lab to playing Candy Crush Sage. You've got good." Daniel looks at me, then lets his eyes rest on the floor.

I look out of the window, disappointed. His lab overlooks the dramatic little mountain known as Arthur's Seat. This unique, volcanic offering sits in the middle of Edinburgh, a lovely burst of greenery for a city centre. High and mighty Edinburgh Castle is also in clear view along with the sprawling countryside around what is actually a small city – barely more than a large town.

"Nice view, isn't it?" says Daniel. "On clear days you can see the North Sea and across the Firth of Forth to the farming lands of Fife."

"It's a splendid sight, which is just as well because you seem to spend much of your day admiring it."

A feeling of sadness is coming over me. It takes a certain mindset to stand in awe of PhDs; to perceive the honour and to respect the effort encoded into those three humdrum letters. Admittedly it is an elitist mindset, but it genuinely embodies a noble aspiration; ultimately, a PhD is driven not only by the desire for personal affirmation but also by the desire to enrich the common stock of human knowledge. A PhD, for me, is some people's idea of a contribution to society; a contribution they would like to make. The desire is infused with a certain humility even if status is also being sought. Other people want to contribute to society in other ways; greater and lesser ways depending purely on how you care to measure it. However, some people simply do not buy into the concept of contributing to society – at all. To my surprise, my boyfriend transpires to be one such non-believer. I want to shout at him, shake him, ask him if he really is such a loser? But we are surrounded by people, by his colleagues.

"You want to go for lunch now?" Daniel asks running his hand through his thick blond hair that always looks like it's recently had a few thousand volts sent through it.

"Sure. You paying?"

It's a cheap shot but the open goal is irresistible. Daniel does not get paid for the time he spends at the university. Daniel does not get paid for the time he spends anywhere upon this Earth. His Dutch scholarship ended and his parents cancelled his monthly allowance in a vain attempt to pressurise him into the adult workforce.

One corner of his lip curls up and his eyes shoot fire at me. "Please Ka, shut up."

He looks away and, in a different tone, says, "You guys come as well?"

The brown-haired girl and the short blond-haired guy both agree and put down their pipettes. Apart from the tension between me and Daniel, it's relaxed and sociable – it is, gulp, normal life. And I realise normal is now weird to me. My telephone rings in my pocket, I see on the display it is Lucy. I invite her to join the impromptu lunch party.

I turn to Daniel's two colleagues. "Hi, I am Karin," I say shaking the child-like hand of the short guy.

"Edward."

"Stacey. Nice to meet you."

Stacey is very skinny, a bit shorter than me and has large blue eyes with long eyelashes. I recall a story Daniel told me the other evening after watching a movie with a few people from his lab.

"Ah, you are the girl with the one-legged dog!"

She smiles widely, "Actually, it has three legs."

"She has an aggressive rabbit as well," Edward chips in.

"Oooh, she isn't aggressive!"

"It is! It bit me the other day as soon as I entered your flat."

"She is just afraid of other people."

Normal banter – how weird…

We pass their office. Daniel has been blessed with a large desk to stare across, and the office has at least three free spaces. There are two black plastic trays hosting ten small cacti each. Next to the trays there are about thirty well-organised low-fat margarine containers filling up a desk. In most of them tiny green leaves are emerging through earth – I presume these are cacti-to-be. In Lab 262 we would no sooner dare start a cacti farm than ram a cacti up our intimate places.

"Hey Vlad, you coming for lunch?" Edward says to a guy in the office.

"That's the guy who wrote the note," Daniel whispers to me too loudly.

Vlad abruptly turns his head in our direction. He has the face of the slighter fatter twin brother of Mr. Bean, with a trimmed circle beard. He looks much less harmful than his note implies.

"What are you telling about me, Daniel?" he enquires, in an Eastern European accent.

"She just asked who wrote that friendly note above your bench."

"Unfortunately, that is necessary here."

He stands up and comes over to shake my hand.

"Vladimir."

"Karin."

I gesture towards Daniel, indicating he is the connection point. "You can score better," says he, completely ignoring that Daniel is standing behind him. "But you need to let your hair grow longer. Like this is nothing."

Wow, that's a bit cheeky, my hair isn't so bad... But 'score better' is hunky dory, I should co-co!...

"I like your honesty and directness," I say.

"Good, most people don't like it. Or simply don't get it."

He adds this comment while patting the back of a long-haired, overweight guy in a track suit. This chap looks misplaced; much too tall for the low desk, too chubby for his office chair and with fingers too fat for the keyboard he is tapping.

"I gave him a print-out with ten eating manner rules, but look at him..."

He uses the guy's shoulders to turn the chair towards me. "Still, Chinese noodles sticking at his T-shirt everywhere."

I fall silent. He just treats this poor guy like a piece of meat we are checking out. I feel embarrassed but the guy doesn't seem at all bothered. "He just doesn't get it," Vlad says despairingly while turning the guy back towards his computer. "At least he won't spoil our appetite now; he has already eaten."

I am watching the scene with wide open eyes. The others also look uncomfortable with the situation, but less surprised. "You can't do that," I say very quietly.

"Let's go," Edward says, slightly pushing me in the direction of the door to avoid any conflict.

The five of us walk to the elevator and wait until it moves the eight floors up to fetch us.

"Sorry, but I really don't think that was very nice what you did there," I say to Vlad, still not having the faintest notion what sort of man I am addressing.

In a very friendly tone, he replies: "You know what is not nice?… To hear him slurping Chinese instant noodle soup next to your ear, soup which he eats with his *hands*. Have you ever seen the consistency of Chinese noodle soup? You can't eat it with your hands, as you just saw from his T-shirt."

"Okay, that is quite disgusting. But still…"

"I already went to my boss three times to ask where he got him from. The last time I asked him if he started recruiting new PhD students at the zoo."

„ Are you recruiting your new PhD students at the Zoo? "

"Is he working well at least?" I ask, thinking he might be a classic eccentric scientist.

Vlad laughs as if I just asked if Katie Price won the Pulitzer.

"He can't use a pipette. I need to show him everything ten times, and he simply does not get it."

"But he must have a degree, right?"

"Yes. He has a bachelor's. God knows where from, somewhere in England, maybe he stole it."

Daniel presses "7" and I wonder why we don't just walk down one flight. When we get out, Stacey says, "I just need to check on Sneezy, I'll join you after."

She presses the "0" and the elevator doors close behind me. *Did she say "Sneezy"?*

Edward explains, "Sneezy is the aggressive rabbit. It's in the car."

"In the car?"

"Yeah, it can't be home alone so she brings it to work every day."

"It stays in the car?"

"Yes." Slightly worried, he adds. "Don't give her the idea to bring it to the office. It's aggressive and it stinks." *She could entertain bringing a rabbit into the office?! This is all so sweet I want to weep for us in Lab 262.*

Lucy is waiting for us at a table near the back of the canteen. We don't need long to decide what to eat as the canteen choices are paltry in that Spartan way beloved of academia. I take a salad with fries and pay for myself and Daniel.

"You are not much of a gentleman, are you?" Vlad comments from behind us in the queue.

"I don't get paid, she does."

"Oh, come on. Get a weekend job at Sainsbury's."

Good man! Well said!

"I am not going to work at Sainsbury's."

"Then you will be a loser without money."

I'm really starting to respect Vlad's blunt, rude, over-assertive, inappropriate and painfully true interventions...

Lucy greets us all with her minimalist smile. It's all you need when you are divinely beautiful. In fact it was probably more than was advisable with Vlad. He pushes Daniel to the side so he can take the seat opposite Lucy. He looks at her as if he has just found a diamond in a dung-heap. Instead of a handshake, he almost kisses the back of her hand but pulls back in the face of her unwelcoming expression. He is watching her like a starving pig staring at a mountain of fresh slop; in that unsexy way which is the speciality of goggle-eyed, spotty adolescents and – sadly – some academics.

"Hello," he says, in a dreamy voice that affects everyone; it is as well that we have not yet eaten.

"Hi," Lucy replies drily.

Her tone really does make it plain that this woman's juices are not flowing, but Vlad the lad will not be deterred, "You have a boyfriend?"

"No."

"You want to go on a date?"

It's impressively direct. I'm worried Lucy finds it disarming. I'm not sure gorgeous creatures like Lucy belong in academia. I don't know where they belong – her beauty sheers her off from everything around her – but it's certainly not in Vlad's bed. Then I think about the weirdo from the Chemistry department she went to the cinema with, 24/7 veggie boy. This Vlad is at least as strange, but less attractive.

Lucy and I exchange a short glance which confirms she knows Vlad is not a person any self-respecting girl will waste her time on. She answers with Vlad-like, dry directness, "No."

A very serious, almost caring expression emerges on Vlad's face and I just know he has a priceless line for such situations. True to form, it's a cracker, "You know, you are actually too old already. You won't have much chance anymore. I can take you. I will buy you nice presents."

Lucy laughs, and the whole table convulses.

"I am not interested, thanks."

"Okay, but I will keep trying. I promise."

We were all taught to be honest, to say what we mean and mean what we say. Perhaps Vlad's parents pushed that lesson a tad too hard. He could get done for harassment in the office and punched in the pub for being a pest. But we academics are too anaemic for either. Is our studious world a refuge for lads like Vlad?

Perhaps strangely, I like Vlad. He is entertaining but, also, his unsparing thoughts and harsh judgements chime with my new thinking in Edinburgh: I fear the only thing blossoming in my life as a PhD student is cynicism.

Chapter 11

Through the cracks round the door I see it is still dark in Lab 262. This is a little eerie because I am not early today. I went to the pub with Lucy yesterday evening, after an eight-mile run round Arthur's Seat, so my legs were doubly reluctant to leave the bed this morning. Still, it's good to get a quiet start. I switch on all the lights, drop off my backpack and turn on the radio to dispel the unexpected silence; Metallica with *One*, a heavy guitar riff to kick off the day – perfect. I drop off my coat and inspect my lab book to remind myself of the exact PCR methods I used during the preceding four days, none of which yielded anything. The PCR product I am looking for is not worth looking for; it could be purchased custom-made. But a few months of my life are far less valuable than a few hundred pounds, in Mark's world anyway. *A couple of degrees up at this temperature cycle, an elongation step that is just a few seconds longer, down a bit there…* it is as intellectually stimulating as optimising the boiling time for an egg. The difference is that a boiled egg is usable. Dutifully, I come up with three new PCR cycles to try.

"Today I will get the fucker," I mumble with my lips closed.

I take the *Burkholderia* DNA out of the freezer, three sets of different primers, two buffers produced by different suppliers and a small tube with polymerase enzyme. Not much enzyme left I note with a nervous glance at our shopping list, but hopefully sufficient for my three PCRs.

I place the tubes in the machine and turn it on. It is 9:00 a.m. when I enter the office so Hanna should walk in any moment. She is very predictable; she works from nine to five – not longer, not shorter, not by a minute. This morning, 'clockwork Hanna' she is unaccountably absent. I start eating the bread roll I bought on the way here and look lazily at a scientific paper

© Springer International Publishing AG 2017
K. Bodewits, *You Must Be* Very *Intelligent*,
DOI 10.1007/978-3-319-59321-0_11

lying on the desk in front of me. Shortly after finishing half of the bread and having absorbed about three words, I look at the clock, almost 9:15 a.m. I am still alone. *Strange. Did I forget something? Is there a meeting or something?*

Slightly nervous I go to the common computer and open my student email account. One new message, from Beverley Alexander, entitled: "Disclosure Scotland". I scroll through the old emails, but there is nothing to suggest I might be missing something important. Also, Lucy did not mention anything about a meeting yesterday evening. I decide this emptiness is down to coincidence and open the single new email.

From the administration office of the School of Chemistry:

Dear Karin,

I have not received your filled-out Disclosure Scotland form for your teaching duties at the university yet. As your classes start this week I urgently need it.

Don't hesitate to drop by my office if you require any assistance.

Best wishes,

Beverley Alexander

What is she writing about? I must have missed it.

It is nearly 10:00 a.m. when I amble along the large hallway to the front of the building, to see that the small lecture theatre is empty and the larger one is slowly filling with undergraduates. They all look as if this is much too early for their biorhythms. I spot none of my colleagues and decide to check the small hidden room behind one of the theatres, where we had a lab meeting the other day. The door is locked. *Where the hell is everyone?*

The door of the administration office is wide open. I knock softly on the doorframe before entering. When both ladies behind their computers look up I step closer to their desks, and catch a pleasant whiff of Chanel perfume. I speak with fake insecurity:

"Hi."

"Hi, how can I help?"

The short, blond-haired woman has a friendly voice and a nice smile.

"Well, you just sent me an email about this Disclosure Scotland thing I need to fill out..." I say, trying to sound like I care deeply about whatever bureaucratic pedantry I have not tended to. She nods, indicating she wants me to continue talking.

"I am not sure I know what it is about. I don't think I received anything."

"Oh, didn't you?" she asks, still friendly but a little edge creeping into her voice now.

She gets a folder from one of the shelves and presents me with a green and white paper. "There it is."

It looks like a very standard official document, the sort that gets filed away and never looked at because it served no real purpose in the first place except as part of an employment scheme for those weird, mentally ill people whose hearts don't sink when they hear the words "rules and regulations."

"What's it for?"

"Everyone teaching at the university needs to fill it out. It is a standard check, to see if you have been convicted of paedophilia for example."

"Paedophilia? Why would that be relevant for my teaching?"

Before she can answer I add, "I am not teaching at primary school."

"No, but some of the students you will be teaching are still seventeen."

What? You cannot be serious? Are you mad? Are you so under-employed you have time to contemplate the ridiculous notion that a seventeen year old who has recently escaped from under his mummy's protective wing and is now partying like crazy would see reason to press charges if a twenty-four year old graced his spunk-encrusted duvet cover with her body?

"Do you think a seventeen year old would mind?"

She looks at me surprised. In all fairness, she is clearly trying to suppress a smile and she barely manages – she is not lost to the devouring paranoia of bureaucratic process just yet – it usually takes a few years for the soul to atrophy completely.

"It is illegal in the UK."

I smile politely, "Oh, okay, I will check their ID first before taking them home."

She smiles and chews her under lip, almost commenting but just not comfortable enough with bureaucratic heresy.

"I'll fill it out and bring it back," I say after a few seconds of silence.

"That's great!"

"Can I ask you something else?"

"Sure."

"There is no one in my lab today, which seems strange to me. Do you have an idea where they could be? Did I miss something of utmost importance?"

"I can have a look?"

She opens a different window on her modern flat screen computer, which means she has better equipment for protecting 17 year old males from having sex with 24 year old females than Lab 262 has for fighting cystic fibrosis. She shakes her head. "No, there is nothing special in the calendar."

"Strange."

"Maybe a lab meeting?" she suggests.

"Not that I'm aware of."

She presses her lips together and lifts her shoulders.

"Well… I'm sorry I can't help."

"No worries. Thanks anyway."

I stride to the lab, this time taking the shortest route. I walk from the front to the back of the building through the long corridor on the second floor which, in the middle, everyone regrets taking. It's the only hallway in the department with a thick brown carpet. The floor beneath is flay, crumbling and just plain not here. It has holes and ridges everywhere. It squeaks like crazy and sometimes springs a little. I speed up to reach a safe base before the whole construct collapses and everything falls down one storey. I open the door of the lab and turn the volume up on the radio, put on gloves and carefully apply masking tape to close the ends of the plastic form we use to make DNA gels. The music makes me happy and I am hopeful that the PCR method I tried this morning will work. Just when I add a few microliters of the poisonous Ethidium Bromide to the hot liquid in the Erlenmeyer flask, I hear a loud sigh behind me. In shock I very nearly drop the hot glass flask but manage to get my shaking fingers under control. I turn around and see Mark stretching up above the refrigerator to switch off the radio.

"You did not hear me entering," he states, looking very annoyed about me being so engaged in my work.

"Eh, no, I didn't."

"No wonder, if you have that so loud! No one can work with that noise."

"Yeah, I was alone, so I thought…"

"Where is everyone?" he is positively barking now.

"I don't know."

He shakes his head. "You don't know."

He repeats this as if I am at fault for not knowing the whereabouts of these other adults who I don't live with. His keys tick on my lab bench. He regards me as if I might have killed the other students and hidden their bodies in the cold-room. Or maybe he looks at me as if he wants to kill me. I don't know. For sure, behind those slightly bulging eyes there lies barely contained fury.

He peers at me for a few seconds then takes a few steps.

"Don't you think you should tidy this up?"

He points at the sink, in which there is half a dozen plastic bottles, some filled with old broth and lumps of pink disinfectant powder, alongside loose props of cottonwood and aluminium foil which we use to close the flasks. Next to the sink there are two Styrofoam ice buckets in which a few, hopefully empty, Eppendorfs float. There are gel-smeared plates everywhere. And there are used paper towels, a bottle of chloroform and dirtied metal spoons. It is, as Mark observed, a mess. But why on Earth is he pretending it is my mess?

"Sure, but it is not really mine," I dare to say.

"No, it's never anyone's!"

He has a point there, though only in a literal, meaningless sense. It is never anyone's mess. In our lab it is almost impossible to find your own mess amid other people's mess because it is a small, congested, demoralised lab – one homogenous mess. However, passive-aggressive has long been Mark's metier with the others, and I suppose I should be grateful this is only the second time he has been openly hostile to me. But it's a surprise attack, I feel ambushed and dazed – better martial my defences… Breath in, breath out…

"I guess the bottles are left there to let the disinfectant do its work before we flush it in the sink," I say, knowing that I can't explain the other mess that way.

Again he shakes his head, puts his keys in his pocket and starts cleaning the bottles himself. I start to clean the glass plates and empty the ice buckets. From the corner of my eye I see the Erlenmeyer flask standing with the hot liquid.

"I really need to pour that gel before it settles."

Mark doesn't comment. Just as I hold the Erlenmeyer at a 45 degree angle to start pouring, Marks asks: "How far are you with LpxA?"

He uses a gruff, demanding tone of voice that seems more befitting of a Kremlin big-wig addressing a lackey than academics working together on scientific research in Scotland's capital city.

Breathe, breathe, bloody breathe! Don't shout back at him. You are above all that and it will get you nowhere.

"I am working on WaaA and LpxC," I reply, fearing this will confirm I am a bad person.

He scowls. "Have you ordered the primers for LpxA yet?"

"No," I say, ever so softly, head bowed.

My knees are getting weak, my heart starts beating at high speed and I want to sink into the ground. *I am sure we agreed I should work on WaaA and LpxC. Why does he suddenly bring up LpxA? Did I misunderstand? No, I didn't!…*

"I want you to work on that now," he snarls.

"Okay," I say like an errant three-year-old who has been caught cutting electric cables.

I'm thinking it would be fitting to clip the heels of my boots together, place two fingers under my nose and extend my right arm at an acute angle, but I'm too scared to wear an expression let alone make a sarcastic gesture.

When the sink looks sort of tidy Mark walks to the cold room. Briefly he glares inside the eight square metres and closes the door again. Without saying a word he swings open the door and storms out of the lab. I sit down and take long, controlled breaths. *That arsehole! Did I do something wrong? No, I bloody didn't.* I feel emotional, about to cry. Maybe I am crying. Or maybe I am just angry. I don't even know. I fetch a pack of Marlboro from my backpack in the office and walk downstairs. I smoke one, then another. Slowly I get myself together. *Don't get angry. There a plenty of people who envy you for having this position. He is just moody, that's all...*

I walk back upstairs and try to concentrate on my PCR sample that now needs to be pipetted into one of the tiny slots of the DNA gel lying in a buffer. Just when I pull the pipette tip out of the slot, with hands that are either trembling from nicotine overload or nerves, my telephone vibrates. A text message from Daniel; *Do you want to go for lunch?*

He really thinks I will buy him another lunch?

I would love to talk to someone about what just happened with Mark, but Daniel probably doesn't have sympathetic ears for me. Ever more we meet each other at awkward angles.

"No," I reply.

I close the box and turn on the power, little bubbles form in the buffer indicating that there is an electric field and the gel is running. Once more I check that I have the cables connected on the right side. It would not be the first time that I have the anode on the wrong end of the gel. I realise I am still a bag of nerves.

I would love to turn the music back on, but I don't dare. *What if he comes in a second time?* Instead, I walk to the office and open the full genome sequence of *Burkholderia* on one of the shared computers, all of which are badly scuffed. I go to the search engine and type "LpxA." No hit, I need to search for it myself, I feel ridiculously harassed. Quinn comes in, says "Hi" and plops himself on one of the chairs. As on most days, the smell of perspiring feet wafts in with him. He takes care of himself in general but, alas, seems to possess only one pair of socks.

He quickly checks something on a computer.

"Mark came in before," I say.

I feel the urge to talk to someone. I need to reduce my stress levels somehow. But Quinn and I never really talk. So far, my only good contact in the lab is Lucy.

To my surprise, he turns his chair around and smiles. He has a lovely smile. It is lively, friendly. "You got the shits, didn't you?"

"Yes!" I say, even more surprised.

He laughs and turns back to his computer.

"How do you know?" I ask.

He shakes his head the same way Mark does, except that Quinn smiles.

"What is it?" I ask.

His smiling is somehow contagious, but at the same time annoying.

He walks to the lab and starts to search for a chemical. He looks my way several times, very playfully, relishing my need for explication. From his secretive smile I sense there is no chance of an explanation any time soon. He is teasing me and I am not going to gratify him. I ask no further questions and go back to work.

A few minutes later, Erico walks in. Finally and mysteriously there seem to be more people arriving. It is lunch time, and it is getting lively here. I look at the sweater Erico wears; slightly too large for his short body. Neutrally, I ask him if I missed something this morning. I have obviously disturbed his concentration but he answers: "No. Why?"

Before I can reply, Quinn walks toward us, still wearing that knowing smile, and says: "Mark gave her the shits."

Now Erico laughs. *What the hell is wrong with these people?*

"You were in early?" Erico asks.

"Well, not that early, but much earlier than you guys," I say, feeling more frustrated.

Now Quinn openly laughs. He regards me pitifully. I'm feeling emotional, desperate, almost like crying. "Can anyone tell me what the hell is going on here?"

"Celtic lost, Ka. Celtic lost."

I look at Erico, no doubt with my blue eyes wide open like a very stupid child.

"That's Mark's football team. Never come in early when Celtic lost."

What the fucking fuckety-fuck?!

Chapter 12

Gorgie Road is teeming with cars and buses when I shove my bike out of the hallway into the street. Over time the faces of the people at the nearby bus stop take on a reassuring familiarity. I am another regular face, part of Edinburgh, or so I tell myself, despite my peripatetic lifestyle and nebulous work.

I cross the street and cycle straight up Henderson Terrace; the toughest part of my journey to the university in this very hilly city. By the time I reach the next incline my leg muscles are at least lukewarm. After this uphill, which lasts about ten minutes, it is payback time when I turn right onto Morningside Road, charge full speed downhill and break to turn left into mysterious Newbattle Terrace. This residential street is stone-walled to the height of about nine feet, with weeping willows hanging over the big coping stones at the top, almost entirely concealing the villas behind; it is expensive privacy rather than ostentatious wealth, but it is striking just the same. Due to the high walls, the street is dark and has its own microclimate; moister and two to five degrees colder than the rest of town. The pothole-rich tarmac only dries for a few days in spring and autumn. Looking at the sky, I think it might actually be sunny today – spring is in the air. As I near the university, the eerie atmosphere disappears slowly – but, in Harry Potter city, never completely. When I glimpse that classic Victorian edifice, the Royal Observatory, perched upon Blackford Hill, I know I will only once more have to stand on my bike pedals to reach King's Buildings campus. The journey was exciting at first. Now it is merely pleasant. Even so, it is sometimes the highlight of my working day.

I park my bike on the small meadow in front of the Joseph Black Building and see an undergrad pacing up and down outside his car, parked

© Springer International Publishing AG 2017
K. Bodewits, *You Must Be Very Intelligent*,
DOI 10.1007/978-3-319-59321-0_12

right at the entrance. Beneath his matted blond hair, his face is liberally decorated with red spots and blotches suggesting chemical research of the ill-advised sort. If he wasn't too young for it, this would be the guy who you think wrote the movie script for *Frankenhooker*. He wears a white T-shirt below a denim jacket, and relatively wide trousers offering an undesirable view of the top of his bum-crack. More and more, he is there when I arrive at the university in the morning. He seems uncomfortable, nervous maybe. The way he checks his trunk over and over again makes me wonder what he is hiding in there – obviously a corpse springs to mind, though it could be a heroin shipment too. The only reason I know that he is indeed a student and not a random freak is that I saw him at the PhD open day when Logan and I presented our lab. He chatted to Logan and I took an immediate, unreasonable and yet passionate dislike to him.

"Don't be friendly to him," I had whispered.

"Why not?"

"Logan, he is a freak!"

"Oh Ka, you can't say such things about other people."

"He has a dead body in his trunk… or at least a dead dog."

"Shoosht… not so loud."

Chills run down my spine and I briefly wonder if it would not be better to turn around and run. *He is such a freak. What is he doing here at this quiet time?*

I quickly pull my student card through the card reader at the front door and type my entrance code. When I hear a short zooming noise, I know the door is unlocked for a few seconds. I am relieved to hear someone whistle in the corridor. *There are more people in the building, and the freak did not follow me.*

The whistling is getting closer. It is the red-haired lady I saw in the chemical stores for the first time a few days after starting my PhD. During that first encounter and during every regular sighting since, she has been whistling the very same tune. She smiles at me and turns abruptly to the left to walk down the hallway to the back of the building. With each step she forces her heels a little too far up as though her tendons, like her jeans, are just too short – a faint homage to Monthy Python's Ministry of Silly Walks, or a weird crossbreed between Heidi and Pipi Longstockings, I can't decide, but it is a jaunt that makes me smile every time.

"Good morning, Babette," I call brightly when I open the door of Lab 262 and see my PhD peer, who is one year ahead of me, violently crushing the keys of the keyboard attached to the HPLC machine.

She responds with frosty silence, as is her wont. For weeks that's been our little ritual for starting the working day. It is not uplifting, it is comically depressing. Despite her not being much of an early morning person, Babette is probably getting up before the rooster crows to avoid colleagues, to avoid human interaction. I have realised that in a busy lab, with students and researchers conducting similar experiments, it is survival of the earliest; leastways if you don't want to crack up waiting for your turn to use some cheap, mediocre equipment. So I come in early, like Babette. It is like an unofficial time-war between her and me – let's see who is in first today, as if it matters a broken Bunsen. Win or lose, Babette detests my presence, but I don't have time to care. I am here to get a job done, I am going to be a doctor…

Briefly, I let my eyes rest on her long, light brownish lanky hair that hangs around an ashen face with unremarkable eyes for a moment. During the dark winter in Edinburgh I hardly noticed the transparency of her skin, but on a bright day like today the network of blood vessels shining through the epidermis is all too visible. By now, I can judge her mental state by the pulse on her temple. And today doesn't seem to be a good day, even by her dreary standards. She is groaning loudly, puffing and making other curious noises expressing displeasure. I should really have arrived before her. *She will badly abuse the precious few pieces of lab equipment we have such that she is the last one to ever use them. Mark really does not seem to have a single penny left to buy us a new keyboard, and definitely not another HPLC.*

I switch on the stereo on top of the −80 degrees freezer. Irish folk music fills the lab puncturing the tension, or so I choose to believe. I open my lab book containing all the notes that seem to have become my sole reason for living. I take a falcon tube with a muddy mass of *E. coli* out of the freezer and slowly start sucking the cells up and down an ice cold buffer with a pipette until a defrosted homogenous mixture has been created. Meanwhile, the falcon tube needs to stand on ice, making the process even slower. If you don't mind watching a washing machine running through a full programme, then you wouldn't mind this job. Alternatively, if you are sane then you will wonder how your life came to this sad pass.

Just as I have the bugs swimming, in walks skinny Barry; the recent recruit to the postdoctoral ranks, funded by that money pot which came and went in the fizzle of a struck match. Babette might be comically depressing, but Barry embodies academic misery without a whiff of light relief. Just watching him is saddening. Every day he drags himself up the hill to campus, shoulders hanging and puppy-eyes pleading to escape this wearying existence. He is already utterly sick of the job he just started. His forehead is furnished with astonishingly deep worry lines; these do not auger well for the ageing process which he has prematurely embarked upon in every sense. It's his second postdoctoral position in academia; a sort of reluctant stumble into an accidental career. Lack of success with other job applications sucked him in – again. He tries, and fails, to hide that he hates it here. I'm delighted he is here: Mark adores stressed and defeated Barry, and thus he has adopted him as his new target for evening monologues.

Today, he is quickly – sort of obsequiously – followed by a spoiled brat, the new project student he is supervising. The contrast between the overly enthusiastic girl, speaking with a posh Cambridge accent, and the disillusioned and depressed Barry is bewildering – like an accident you have to watch. They go into the office to deposit their coats and bags and return to start the procedure I had just followed. Theoretically we could join forces, not to all repeat the same, mostly pointless, exercises, but that is unthinkable in Lab 262. This is a work environment where dislike and suspicion are cooked up as efficiently as any compound in a test tube. By now, I credit Mark with this atmosphere, these people, the draining ennui and all the time-wasting: this toxic mix will, I suspect, be his biggest contribution to the world of science. But, hey, I'm going to be a doctor…

I watch the couple for a while from the other side of the bench. She pushes the pipette into the frozen cells and pours a bit of buffer into the tube. I wonder if her tank top, presenting the upper halves of her spotty boobs, disturbs Barry as he patiently shows her how to suck their *E. coli* mixture up and down, but then I decide that Barry probably has got other things to worry about. Though accustomed to the *Theatre Bizarre* mornings, I feel uncomfortable and am happy when Lucy, Hanna and Logan slope in and we move away from Act I; things start to "normalise."

At 9:30 a.m. Mark comes in and a deep sigh goes through the office. Barry rolls his eyes the same way Lucy does sometimes. He knows too well that Mark will capture him like prey, and half an hour of his life will pass, never to be seen again. Despite me being pretty sure Mark will ignore my presence, I feel my underbelly contracting and a weird tingle shoots from my elbows to my fingers. I know my limbs will feel slightly shaky until he leaves. Since Barry's arrival, I have got into the habit of sneaking out for a cigarette as soon as Mark starts droning at his victim. Today is no different.

While walking downstairs, I hear loud footsteps behind me. It is Babette, also on her way to smoke a cigarette. Also escaping Mark, I suppose. Babette can't walk, she can only stomp like an elephant. It wouldn't surprise me if she shatters her tibia one day by way of this thunderous exertion.

We opt for the same nearby exit, ignore the non-smoking signs and suck on our fags next to a container with empty chemical bottles. It being a day of the month, Babette doesn't look like she is up for a chat. Neither am I.

It is a pointless exercise; putting in the effort of breaking breath with a grumpy person who feels put upon if you say anything to her.

I walk back upstairs, and see Mark still has Barry in his grip. As I pass them, Mark abruptly turns round, facing me. "You checked your email?" he asks in his typically demanding tone. *The carrier pigeon has not arrived yet to deliver our daily basket of emails to Lab 262...* Already he has an expression on his face that no matter what my answer will be, he will be disappointed by it.

"Not yet, no." *Wrong answer.*

Mark presses his lips together and sighs through his nose. *Of course, wrong answer. Do I now endure a rant about when in the day I actually should read emails, or about the content of the email he sent? Just bring it on, dude!...*

For a moment it looks as if he is actually about to shout, but he sighs another time instead, to get himself together. "Prof. Raetz agreed to send some of the antibiotic he synthesised. It's precious material, so be careful with it. You can send your LpxC plasmids to the US, to have his researchers assay them for us, this week," he says, much more enthusiastically.

I fall silent. In terms of LPS research, Raetz's is the most famous lab in the world. He is the world's leading researcher on the pathway I am working on. In fact, the pathway is even called after him, "The Raetz Pathway." He was the first researcher to publish every single step of it – he drew it and he mapped it. He is a Big Cheese and it rather amazes me that he is giving Mark – a nobody in the field – any attention. I guess the attention must be based upon Mark boasting about results which do not, as such, exist yet.

"Isn't that great?" he says, now almost manically, like he just unwrapped his Playmobil Pirate Boat to play with back in his office. *What kind of drug blend is he taking?*

I take a few seconds to think about an answer. "Yes, it is great... But, we don't have the plasmids yet." *And we are bloody far from having them!*

"Then... make them!" *Of course... Bippity boppity akawaka alakazoo dippity doo and boom it's done!*

"I have been on this project for eight months, and so far it mysteriously didn't work."

"I will help you." *Oh please, don't help me! I bet you don't even know how to turn the machine on and off.* "We must get that done!"

"I'll do my best."

He slaps me lightly on my shoulder, smiles and walks out of the lab.

Hectically, I start digging through the freezer to get out all the stuff needed to start a PCR. As our PCR machines are of a similar vintage to the autoclave, I run downstairs to ask another group if I can use theirs. Thankfully, I can.

Around midday, I'm hanging in the office waiting for Lucy to go for lunch. I am resting my head on the table. It feels like it is seven in the evening. The whole plasmid thing now needs to be top priority, just when I was fully engaged on a different project; the one Mark had told me last week to focus on. As we walk to KB House and climb the stairs to the canteen, Lucy and I chat, like friends, and this somehow feels like life-enhancing therapy today. We both decide to opt for a cheeseburger with potato wedges, buy an overpriced coffee to help digestion and forty-five minutes later return to the lab.

In the lab we follow the same pattern as this morning. It is Babette first on the machines, then me (running the plasmid experiments in between), then our dear tear-inducing Barry. I am quietly waiting for my turn and watching Babette from a distance. Her blood pressure has dropped a fair bit during the last few hours but the compressed eyebrows, together with that aural punctuation of odd and disturbing noises, indicate that she is angry or grumpy; in other words, back at her default setting – in a cold war against the world and against herself.

While still waiting for her to finish, Mark enters the lab chatting with a young, Mediterranean-looking man I have seen a few times before in the department, but never in our lab. This guy is a group leader I believe. Mark is making large gestures as he talks. He bins some used paper towels that lay spread over our work bench, as if that might disperse the stifling mood of abject misery which forever hangs in the air of this lab like a stale gas.

"Hanna here?" Mark asks.

"No, she's at lunch."

He shakes his head in disbelief, as if lunch at lunchtime were a bridge too far. Outraged incredulity is his favoured expression when he wants to spread bad feeling among lab inmates. Still, I'm surprised. Normally he doesn't mind us taking breaks. He never joins us for lunch, but during coffee time he is mostly there, resting his ass on one of the dirty couches of KB House – holding a monologue about football or research. The first few times I tried to contribute to the conversation, but it was made very plain that I am not supposed to talk. Only Erico can talk, and now Barry as well I suspect, but he doesn't want to. Erico is able to make Mark laugh with supposedly engaging stories, served up with Italian style. The rest of us are apparently bereft of charm. Our role is to nod and listen. Over the months, these breaks – which are harder work than our actual work – get to me; I am a number, selected for this PhD position based on my CV and letters of recommendation. The actual human presence behind the documents seems to be of no consequence whatsoever.

"You happen to know where she keeps the Sal-1 LPS?" he asks.

"No, I don't."

Oh god, he looks so annoyed, so I nervously add, "But if you urgently need some, you can have some of mine."

"You have Sal-1 LPS as well?" he asks, incredulously of course.

"Yes. Hanna just happened to show me last week how to isolate LPS."

I didn't know if it would actually ever be worth something for my PhD, but after spending weeks unsuccessfully repeating my own experiments, I needed a break from the constant frustration. I had joined Hanna on a trip to the hospital, where she showed me yet another experiment. The LPS came out of it.

When I get the falcon tube filled with a threadlike substance out of the freezer, he actually compliments me, "Well done." *Wow. Which leadership course did you take?*

I hand it to the Mediterranean man.

"Shall I take some out?" he asks in a warm, soft tone with a slight accent that I cannot place.

"Actually, you can take it all. I'm not sure I will ever need it. I just isolated it to learn the technique, nothing more."

"Okay, thank you very much," he says looking at the run-of-the-mill material in the tube as if he has never seen anything quite like it before.

"Is it pure?"

"It needs another thirty minutes of ultracentrifuge after suspending it. I can do it if you wish."

"That's very helpful, but I can ask one of my people to spin it."

He thanks me another time before walking in the direction of the door. Neither man sees any need to share with me why they require it. And I don't ask. *I know my place.*

As they walk out of the lab I catch a glimpse of the undergraduate Frankenhooker freak, walking through the corridor right past our lab. With his head slightly bent to the front, he quickly peeks in, his grey-blueish eyes flicking around. I feel the hairs on my arm stand up... *What a freak!...*

Or is this place driving me nuts?

Chapter 13

"But… but… Hanna showed me what to do…" I stutter.

"No! Hanna's name is not going to be on there!" Mark speaks firmly, pointing at the manuscript of the Mediterranean guy on his desktop. Four authors are being listed, and I am one of them.

"But… eh… but…" *Hanna will kill me! Hopefully quick and pain-free, but if I am unlucky she will poison me with her huge stock of the highly toxic LPS.*

"N.O." He spells the letters out staring hard at me.

I'm at a loss. I look around his messy, small office, not sure if I can or should give it another shot. I want Hanna's name on the paper, desperately. I need to believe there is justice in the academic world, that work is being credited to the right persons. I need to believe I am not an intellectual thief. I'm stalling for time, looking round as if for inspiration. There is a desk, a small table, an armchair and a few bookshelves. High stacks of scientific papers fill virtually all available space. There are basic chemistry books on the shelves which Mark probably uses as reference works for his undergrad teaching duties. Despairing and conflicted, I look to the window, which overlooks the small courtyard of the chemistry building, which no one is supposed to enter. It's all so dismal. The sun is shining outside but this room is as gloomy as a room with a window can ever be, like a cell. It feels fitting. I vaguely ponder if it's even gloomier one floor down. Mark is waiting for me to accept defeat and exit stage left or, more aptly, to fuck off.

Eventually he strikes a note of finality; dismisses me with curt instruction, not unlike a bored tin-pot dictator confidently demanding subservience. "Send me the exact materials and a description of the methods you used, tonight."

© Springer International Publishing AG 2017
K. Bodewits, *You Must Be Very Intelligent*,
DOI 10.1007/978-3-319-59321-0_13

At your service, Sir.

"What format?"

"Word."

"I mean… for which journal?"

My voice is trembling, I've surrendered. And it's wrong; me, him, what is happening, it's all plain wrong, some things are just *wrong!*

"*Angewandte Chemie.*"

I swallow.

"Be happy with it. It's a good journal. It's good for your career. And you didn't need to do much for it." *Very little indeed.*

Tragically, this is a rare moment during life under Mark's auspices when one should be happy. It's not *Nature*, it's not *Science*, but it's damn good! *Angewandte Chemie* [Applied Chemistry] is one of the top three chemistry journals, definitely enough to wet the pants of most chemists. I should be thrilled to appear there, even as second author. Instead, this being Mark's world, which is all-too-typical of modern academia, I feel the ground sink beneath me. I feel shame. And I see fragile relations with my long-suffering fellow inmates of Lab 262 evaporate like meaning in bong fumes. *Never, ever, ever is Hanna going to show me anything again. And I don't blame her.*

With heavy legs I step over two trees worth of paper and leave Mark's office. I know I should be kissing the sky, but I feel a pressure on my chest, and a headache is beating its arrival.

Hanna wasn't best pleased two weeks ago, when I nonchalantly told her that Mark had come in with the other guy to ask for her LPS. I told her Mark had reacted unreasonably, as if her going for lunch had been beyond the pale. I saw her eyes enlarging behind the thick red glasses, her jaw falling down. Her face did not express "Mark-is-such-an-idiot" in her usual way; rather, it suggested I had done something wrong. After a few seconds of glaring at me as if I just told her I am dating her dad, she sneered, uncharacteristically, "If there is a paper coming out it should have my name on it!"

Her tone did not suggest a woman inclined to mercy.

This was probably the first time it hit home to me just how fierce academic competition can be. How such an innocent action could lead to an interpersonal drama; after all, it is the publication record that counts. I had heard poignant stories before; like the lab book of a PhD student from the Johnson group mysteriously disappearing the previous Christmas. The data therein reappeared in a published paper, under her co-worker's name, with no mention of her anywhere. It was shocking, brazen even, yet life carried on with a "shit happens" shrug from all observers.

"Don't worry about it. It is too hypothetical to even discuss such a daft event. How big is the chance that anything will come out?"

I managed to calm her down. She soon agreed it wasn't worth discussing. Such matters seemed of no relevance to us in Lab 262, because our output, as far as the trained eye could see, suggested there would *never* be anything worth arguing about when it came to publications.

Yet it came to pass, and with unfeasible haste. The paper had surely been written already? *WTF?!*

Presumably the Mediterranean guy is a "real researcher" in a proper lab with proper equipment. Maybe he had been actively recruited by the university and is actually one of the rising stars I had hoped Mark was. Mark got a permanent lectureship at an early stage in his career and, I thought, should therefore be a bright spark in the department. But I have since learned that those open-ended tenure tracks are not as prestigious as they sound. People like Mark are being promoted under the internal career advancement system which can drearily and inexorably elevate B-class scientists to meaningless heights. It is commonly understood that such career paths are much beloved by the slovenly and useless. During my time in Edinburgh I will see Mark muddle through the system and witness his elevation from Lecturer, to Senior Lecturer and – shortly after – to Reader. I dare say this trajectory disgruntles Mark in the long run; no title thus gained can obscure his status as threadbare underpants, a research scientist promoted purely and obviously for time served and boxes ticked. Ouch, indeed…

But the Mediterranean must be unconnected to the "automatic career advancement system" of the University of Edinburgh. I might have met the real deal, an actual real researcher, achieving his ambition, and just making a career stop-over in Edinburgh, later to be actively recruited for a full professorship somewhere else. His ambition and presumed passion for science is possibly very impressive, and his political skills downright awesome. So many can only plod, so few can only run…

I go back to the lab, and find Hanna writing at one of the desks. As she will find out sooner or later, I feel I better tell her myself about the publication, but not now. I'll work up the courage later… Nervously I check my lab book to ensure I have all the necessary data and protocols to fulfil Mark's stipulations; in order to send him the materials and methods section this evening. I slip the blue book into my bag, quickly finish my lab work and leave much earlier than normal to go home. I feel reduced. I *am* reduced.

As I open the door of the apartment, I hear Daniel walking towards me. He finished his project in early March and has been at home since, for many weeks now. He is applying for jobs, allegedly, though very stealthily it would seem as there is no evidence of said activity.

"A couple of colleagues are coming for dinner," he announces excitedly, before I have even taken my coat off.

I let my bag drop on the old wooden floor and close the front door. "Since when do unemployed people have colleagues?"

Daniel throws his head back and rests his eyes on the roof as if praying for relief from my palpable disappointment in him. He looks so much less happy than when I entered. I feel sorry for him which, in the long run, is probably far more offensive than my sarcasm.

In a friendlier tone, I ask: "Are you going to tell me who's coming?"

He pauses for a few seconds and says, "Vlad and Edward."

Vlad has come to our flat a few times since our first encounter at the biology department. Though he drinks not a droplet of alcohol, he entertains me with his views on the University of Edinburgh, the biology department and life in general. Nothing is censored, he has no diplomatic filter, and he has angles on everything. "Cool, I like those guys."

Daniel is relieved.

"I won't have much time to join you, though. I need to finish something for Mark this evening."

"Of course you do," he says, slightly irritated.

I lean against the kitchen and tell him about the text for the paper I need to finish and the story with Hanna.

"Wow, that's pretty fucked."

Daniel isn't known for analysing everything to death, or for his inner need to solve complex problems, but perhaps that renders him a convenient collocutor. He listens and says supportive words, rather than annoying you with all sorts of solutions you don't want to hear. I'm hard on him at times. He is a good man, a good person to share a glass of wine with, maybe even two…

"It is."

I go to the living room and open my laptop. Daniel follows me and hangs in the doorway.

"Ka? Why are you always working?"

Not this rubbish, not now, dude.

"I just want to do a good PhD. I want to publish."

I do not say that for weeks I have spent most evenings either in the lab or in the pub in order to minimise the time at home with him. I do not say that I can't listen to his stories, his dreams of becoming rich and what "special offer" rubbish he bought from my tight PhD stipend at Lidl's. I do not say that I can't bear to see him run his hand through his messy hair once more while he smokes my cigarettes at the kitchen table. I do not say that I am sick of coming home to find the house in an absolute mess when he has been at home the whole day. I do not say that I simply can't handle sitting in the same room as him anymore. I do not say our relationship has died, that I feel embarrassed by his ambition. I do not want to let that devouring sadness into the room.

"Okay," he says in a way that at least sounds satisfied, and he goes back to the kitchen while I start writing.

I read a random article recently published in *Angewandte Chemie* to check the house style. *Champions League Chemistry compared to the village football we play in Lab 262*. I start typing exactly what I did in the lab so that every researcher who would want to repeat the research could do so, even though I strongly doubt any ever will. There are a few things to look up, but I finish the text much quicker than I expect, and conclude that if I ever write a book after my PhD, I will opt for a cook book; surely the easiest way to fill up pages quickly. I take the manuscript Mark handed me and try reading it, but it deals with matters far from my research field. I don't totally understand what it is all about, I can't even decide if it's useful or just throwing a hell of a lot of pseudo-light on non-existent problems. Luckily for the authors I am not judging it on its significance to the world. For me it doesn't matter. I stuff the paper back into my backpack, press the send button for my email to Mark, close my computer and join the guys in the kitchen.

The next day I come in shortly before 9:00 a.m. and nervously stare over the bottle graveyard below while waiting for Hanna. I hold fire until she has taken her coat off and settled into the seat she occupies for the first ten minutes of every working day to plan her experiments.

"Hanna?"

"Yes."

"Can I maybe talk to you for a second?"

"Sure."

"You remember the LPS we talked about two weeks ago? With this young guy who came to ask for it? It will get published after all."

She crosses her arms and looks at me with her big brown eyes, awaiting whatever is coming and fearing the worst. She doesn't ask what the research is about. It doesn't matter. All that counts is authorship. "I really tried. But Mark does not want to add your name to the paper."

With a tone of voice that is somewhere between shouting and squeaking, she blurts at me, "I knew this would happen!... This is so unfair!..."

I opt for a silent reply, let her get it out, but I can think of nothing to say anyway.

She shakes her head. "What did Mark say?"

I actually stutter, "Not much… that I did the hands-on work and… and therefore I should be co-author. I tried twice… but he wasn't up for discussing it… not at all…"

"I stood next to you and told you exactly what to do!"

She is now shouting more than squeaking.

"I know."

"Did you tell him?"

"Yes! I did."

She squeezes her hands into fists and takes a deep breath.

"I'll talk to him!"

She strides out of the lab, leaving me feeling relieved, for having told her, but also feeling about two centimetres tall for being party to shoddy deceit and needless callousness.

I guess Mark isn't in his office yet, but as it is semester time and he is teaching an undergrad course, she might be lucky and catch him this early. It doesn't take long for Hanna to come back.

"He will add me," Hanna states, obviously relieved.

"Oh cool." *Thank god.*

"You will be second, and I will be third author. It should have been the other way around, but I can live with that," she says, irritated.

"Okay," I reply, neutral.

"He is such a bastard. I can't believe he did not want to add me." *Hallelujah! Mark is the enemy now…*

"Yep. Luckily it's all fine now, we're both authors on an article we didn't do much for." *And we both don't understand.*

Hanna smiles.

"Where is it being submitted to?"

"*Angewandte Chemie.*"

Caught by surprise, Hanna slaps her hand in front of her mouth.

"AWESOME."

Carrying notebooks, which we hardly ever write a word in, Lucy, Logan and I are walking to the small lecture theatre for the obligatory weekly seminar of the bio- and organic chemistry section.

"What was that with Hanna all about?" Lucy asks concerned.

"Just a stupid thing about the authorship on a paper," I say, and tell them an abbreviated version.

Babette storms past us, also holding a note book and a pen, not saying a word. "I can imagine Hanna didn't like that too much," Logan says, while all three of us sit down in one of the middle rows of the theatre.

"Me too," I say.

There is a young, tall, handsome blond guy uploading his presentation on the computer.

"Boring presentation or not, we've got something nice to watch for the next thirty minutes," I state.

Logan hisses.

"I hope you're not being serious," Lucy says, rolling her eyes.

"He's good-looking, isn't he?"

"Have you noticed he doesn't even wear shoes?"

"That makes him unique, mysterious somehow."

"We have different taste, Ka. I do not find guys who present in front of the whole department in woollen socks hot."

"Check his torso! And the chest hair sticking out of his shirt. Seriously, he could wear a wrap-over vest and still be hot."

"I don't want to imagine this guy in a vest," says Logan. "It's just wrong! Could we please change topics?"

"In a sec Logan, we are not finished yet... Ka, are you joking? He could star in *Planet of the Apes.*"

"At least he doesn't look like he would slink off with his tail between his legs if we encountered a bear in the forest."

"Yes, Ka, but we're in Edinburgh, not Kamchatka. What if your next hometown is Paris... then you would walk along the Champs-Élysées with your monkey, because he asked you out, didn't he, to go eat some bananas..."

Logan is putting his palms over his ears. The blond guy positions himself, confidently, on his thick woollen socks, in the middle of the room, and starts to talk with a heavy, hoarse voice. *Why do I find him so charming?*

Within a few minutes I forget about the stupid paper, my research, Daniel... and everything else that drives me nuts. For a moment, I am just here and now, and open to enjoying life again.

Looking back I was profoundly relieved the meaningless authorship farce was over. It never occurred to me that it wasn't over and, in a way, still isn't. Because of the name of the journal this sordid little farce, dealing with matters far, far away from my research field, is a highlight on my curriculum vitae to this day, and possibly for the rest of my life.

Chapter 14

It is 11:55 a.m. when I open the door with the yellow biohazard sign. For the first time in months I notice a strong *E. coli* reek filtering into my nose. One of our MSc project students, a blond girl with long curly hair and a pleasing smile, seems to be dancing the day away at a bench. With her headphones on she doesn't hear me enter so I have time to observe that despite her ample frame, she is a rather groovy mover at least in this unselfconscious state. She started about three months ago, working in our lab just a few hours a week. She confidently ploughs her own furrow, uninterested in the lab's inmates to an almost rude extent. However, I came to like her somehow, maybe as she conveys the impression that she is one of the most intelligent fishes in the student pool. When she sees me passing the bench on my way to the office she takes one ear-piece out – just one, no need to take too much interest in this loser lab.

"Finally there is someone!" she says excited, but with a clearly annoyed undertone.

"Good to see you too Elli, how're you doing?"

"Great, I'm stuck with the UV-vis spectrometer. Do you know how to change the wavelength?"

I walk to the machine and show her how to change the wavelength. "Where is everyone?" she asks while I play around with settings on the software.

"A few of us are at this global endotoxin conference, in Appleton Tower, downtown. Lucy is in St Andrews. Logan is probably running a practical class for undergrads, and the others doubtless enjoy holiday class II."

"Holiday class II?"

"Yep, that is when Mark is not here."

"Whereabouts is he?"

© Springer International Publishing AG 2017
K. Bodewits, *You Must Be* Very *Intelligent*,
DOI 10.1007/978-3-319-59321-0_14

"At the conference as well."

"How is the conference?"

I shrug, "It's okay. Much of it isn't my thing but Mark wants me to sit through it all. I just sneaked out to work on the plasmids that need sending to the US."

"You've still not finished them?"

"Nope. It just doesn't seem to work. This US dude, Prof. Raetz, is at the conference actually."

"That is sooo cool! You talked to him?" She talks as if Raetz is a Hollywood celebrity.

"No. He's a big shot all right. Mark tried to talk to him, but Raetz is busy with big shot friends… Hanna gave a very good talk."

"At the conference?"

"Yup, she got selected together with two other PhD students to give an oral presentation. It's a kind of competition."

Elli laughs, "*Britain's Got Talent* for scientists… Did Hanna win?"

"We will find out this evening after the conference dinner. It would surprise me if she didn't win. The other two were bone dry."

Hanna had the audience under her dainty thumb the moment she walked on stage this morning, big brown eyes almost crying as she performed her opening. In careful cadences accompanied by eloquent gestures, she expressed her sweet passion, "There are people dying out there… unnecessarily so! And I have it as my target to save them…"

It was sort of unbeatable; half-parody yet genuinely moving in its own way. Mark, being joyless and bereft of irony, hated it. It seemed he was ashamed of having a PhD student who suffered from emotional incontinence and who did not, like the other speakers, present her work as lists of dry facts "no one apart from myself and maybe my supervisor really understands."

"That's great!"

"Yeah."

"Will you be jealous if she wins?"

I think about this for a few seconds… about how gutted and disappointed I had felt that I had not been selected after carefully phrasing and re-phrasing my abstract. Hours went into the pointless activity. I feel like being honest with Elli.

"A bit, I guess."

She regards me pitifully as if I just told her that my granny died. Now I think of me pitifully too, so I hastily add: "But she's two years further into her PhD so it doesn't hurt too much."

I drop my bag under one of the five desks currently shared between seven PhD students, a postdoc and four part-time undergrad students. It is a recent improvement owing to Bubblegum-Bobline and Diet-Coke-Girl being away preparing for their PhD defences. When I return to the lab Elli is really shaking her bootie now in front of the UV-vis spectrometer. I take one of the Styrofoam containers from the sink and walk to the ice machine. I am hanging as far as possible into the tank that would easily fit two human bodies, keeping my centre of gravity just on the right side when the engine spits out a round of ice flakes. I get the full load over me and regret that I didn't put on my lab coat. I walk back to the lab, where Elli again takes one of the ear-pieces out to enquire how to adjust a setting on the machine.

"You do realise that Mark doesn't like people wearing headphones in the lab, don't you?" I ask while fiddling around with the software.

"Doesn't he?" she replies, conveying the impression that this concerns her about as much as the weather in Ulan Bator.

"No. He told me off the other day."

"Oh well," says she lifting her hands to shoulder height, trying to be conversational but making it clear that hearing about Mark's stance on headphones for a second time has actually managed to reduce her interest in the subject.

"I will be out of here after tomorrow anyway."

"Oh really? Tomorrow is your last day? Already?"

"Yeah."

"Are you coming back for your PhD?"

"No!" She laughs almost hysterically. "Mark did ask me, but I am not crazy."

"You don't want to continue studying?"

"Oh, I like science as such. BUT... this place is just grim; Mark being unpredictable and stressed all the time about... whatever. Barry dragging himself up the hill to campus, Babette being ready for mental asylum. And you? ...and the others? You don't even have a desk! And the research coming from here isn't anything fancy... Maybe I will change my mind one day, but I don't see *this* as my future..."

Stop, stop, stop! I want to scream. *We all know that we are working for a B-class scientist in a lab that looks more like a storage dump for a local antique museum than a place where people actually work. We know we aren't doing any great research; not even anything of interest to the fanatical freaks who spend their time jizzing over obscure articles. And, believe me, I know too well that I don't have a desk! But at least I am still trying to paint this rosy, bright picture for myself to believe that I can make it work. How else do you think we refrain from slitting our wrists in labs like this? Stop it! Stupid Elli.*

I want to change the settings on the UV-vis back to "unworkable" and spit in her face for uttering those heresies, but instead I say, "I understand your decision."

I stare at the screen of the UV-vis, feeling Elli's eyes on my back.

"What's your plan for the future then?" I ask with a bright voice.

"I'm going to Australia!"

"Backpacking?"

"No, not really. I got myself a job there."

"Wow. What's the job?"

"I'm going to shoot kangaroos, initially for a year."

"Shoot kangaroos?"

She nods, places the ear-piece back in and presses a button on the iPod mini clipped onto her short tartan skirt which now sways with her majestic hips. I stare into the ice bucket, lift up the Eppendorf with competent cells, and mumble to myself, "You look ready for a short heat shock, my friends."

I place the Eppendorf in one of the slots of a 50 degree block. This is not as dramatic as shooting wild animals in the Australian Outback, but martial enough to kill the day.

„PhD my ass! I am going to shoot kangaroos…"

I hurry back to Appleton Tower as soon as I've spread my *E. coli* cells on an agar plate. I hope Mark has been too busy to miss me. Of course he has better things to do at a conference than look after his temporarily lost PhD students. I quietly sneak in at the back of the lecture theatre without being noticed and no one comments on my temporary absence.

I headed home just after six, threw a handful of job ads on Daniel's desk, and now, less than an hour later, I walk from my flat to the conference dinner; north along Gorgie Road in the direction of Haymarket, passing the tiny alternative video rental shop which is run by two lovely middle-aged men whose development terminated during their teens. During my lonely early weeks in Edinburgh I was a habitué; having a chat and a smoke with the owners whenever I passed. But, having cultivated a semblance of a life and a research project that needs attention, I have deserted these beautiful lost souls. As I am well in time tonight, I pop in to say hello. There is just one customer; a funny looking fellow digging his way through the movies in agitated fashion, clearly distressed at being unable to find the dubious masterpiece he seeks. He sees me, watches me for a couple of seconds too long, drops to his knees and asks me to marry him. *Yes, I would love to see the expression on my dad's face when I bring a chap like you to the next family event.*

"Excellent idea, we could do the celebration in Port O' Leith," I reply.

"You're saying *yes?*"

He comes a step too close. "Oh Charlie, leave her alone," says Keve, the shop owner. Charlie backs off instantly and leaves the shop – it was all too devastating for him.

"Drugs," Keve states, nodding at Charlie's back. "Anyway, how're you doing, flower?"

Every time Keve speaks slivers of saliva fly through the air. He is missing at least three front teeth and, from what I can see, molars are in short supply too.

"Fine thanks. Yourself?"

"Good! Haven't seen you for ages!"

"I don't have so much need to rent movies anymore, seem to have some sort of social life now."

"Aaah, you made some friends? That is good, flower… that is good. How's your boyfriend? What's his name again?"

"Daniel. He's fine. In his own world. Looking for a job."

"Tough times for jobs."

"It surely is. But I am not over-impressed with his determination to find one."

"Too comfy for him at home?"

"Apparently so."

"Stop paying the heating bill. That will get him going."

This light-hearted remark gets me thinking... I am hardly at home, I wouldn't suffer too much. Keve is clearly a man of the world when it comes to indolent types.

"Maybe I could give it a shot in autumn? Just now it won't have much effect."

"Breadcrumbs on his side of the bed? Or give him blue balls?"

"Blue balls? Like kicking him?"

Keve laughs, releasing even more saliva than when he talks.

"No, don't kick him to make them blue! Never mind. You going anywhere, flower? You look dressed up."

It's weird but I feel more cared for in here than I do at home or in the lab.

"Yeah, conference dinner in the Hilton hotel."

"That shithole on Grosvenor Street?"

"Is it that bad?"

"Used to be. They refurbished it recently. But I don't care for posh shite, refurbished or not."

No, not your bag really, full sets of teeth and egghead preciousness, but I appreciate your tricks for shifting couch potatoes, you're an expert in your field...

I move towards the door, "I have to hurry now."

"Yeah, nice to see you again, flower. Give your boyfriend a kick up the arse!"

Oh I think I will! Or at least I really think I might make his arse cold...

At Haymarket there are a couple of drunks lying on the pavement in front of Ryries Bar. As it is Friday evening in Scotland's capital, it would have been more surprising if there were no drunks languishing thus. Just across the main intersection and into Grosvenor Street is the Hilton. The street is residential, no bars and restaurants. It's a strange location. The ginormous shed with the gaudy neon lights reading Hilton seems to sort of violate the sedate surroundings, and I wonder why permission for a hotel chain was granted here.

One of the doormen welcomes me and relieves me of my coat as if I were royalty. I enter a large hall called the Roseberry Suite. It is at least as posh as the name suggests. Round tables are decorated with white table cloths, gold girandoles and carefully folded serviettes. They are positioned around a polished, wooden dance floor. The walls and roof have a yellowish colour and built-in spot lights to enhance their warm texture. Everyone is already sitting at a table and I am one of the last to arrive. Slightly uncomfortable, I move through the seated mass to find the right table. It is only a few seconds walk but feels akin to a catwalk, being watched by hundreds of eyes. I feel out of place but I am probably not the only one feeling a tad iffy.

This is how to get scientists out of their comfort zone. Not only would most scientists agree with me that folding napkins into intricate shapes is a complete waste of energy, but most of the people in this room could, privately, never ever afford such refinements, not on a university salary. I'm sure most guests would feel much more at home in a burger joint. But that would make me feel much less important than I feel now. And yes I do feel important being a guest in such a room, funded by the pharmaceutical industry. It is as if we are playing in the big league, and I am part of a winning team. This is a stopover on the road on to the Nobel... *It's hilarious, the dreams we indulge to keep us going...*

I sit down next to Hanna and a waiter hands me a glass of prosecco. The voice of the dapper organiser booms through the state-of-the-art PA to announce the winners of various prizes: "This year's winner of the Young Investigator Award is Hanna Blom!"

Jubilation resounds around the table; through me, Mark, James, Hanna, Brian, Erico and even two strangers who couldn't find seats anywhere else. Hanna gets out of her chair and slaps her hand in front of her mouth with expertise that Gwyneth Paltrow would envy. The power to imbue speech and gesture with emotion is a power Hanna has in spades. It is a power Mark seems to despise. It is also, arguably, the power that won her this award.

When Hanna stands on the wooden floor to shake hands and accept the award, Mark pokes me in my side, "Next year, you!"

"Sure!" I reply.

Next year's conference is in the US and I know Mark would never fund me for a journey beyond the street corner. Still, this was a Prize; not the Nobel, and not me winning it. But I was there and happy for Hanna. It's a small step in the right direction, I somehow manage to believe...

Soon after the ceremony a five course dinner is dished up and I feel properly blown up to bursting point after just the third course.

A young sporty guy in pink shirt and with some sort of Dali moustache comes to our table and kneels down next to my seat.

"Care for a dance?" he asks with a low voice and a smile that shifts the pointy ends of his moustache almost to his cheek bones. *Wow, I've been upgraded from a drug addict in a video shop to a founding father of the retro movement.*

I look at the dance floor where several scientists are actively proving that dancing is not their forte. "I'm not much of a dancer," I say.

"Me neither, but I'll have a crack if you will."

"How sweet – willing to reveal your lack of talent if I reveal mine."

He looks me in the eye. I feel myself blushing. But he's a good sport, "Drink that up and we'll get on the floor!"

He extends his hand graciously and I drink up, stand and follow him to the dance floor.

"Are you more into tango, salsa or ballroom?" he asks, wrapping his arm casually around my back.

"Eh?... Freestyle?"

"I can do that."

Much too fast we move from one side of the ballroom to the other, not very skilfully and with precious little connection to the music playing. At the end of the song we pause, to my surprise.

"It's lovely here, isn't it?" he says.

"Bit posh."

"Glad you said it. Are you enjoying yourself anyway?"

"This part of it, yes."

"That is sweet of you," he says with a mock grandeur that goes well with his over-arty moustache. "Another dance, young lady? We seemed to be getting the hang of it..."

"Well, we scared most people out of our way as we rolled along blindly like a tsunami. Okay, let's do it again."

That was my posh night on big pharma coffers; being part of a double act, a comic liability charging around the dance floor for hours. It was fun because Retro-King was a gentleman. It was also desultory and peripatetic, not just for me but, I suspect, for all the PhD paupers present; all glimpsing a luxurious lifestyle which has nothing to do with our likely fate.

Around midnight I finally fall into the taxi James had insisted upon. On Gorgie Road I see the light in the kitchen is still on; Daniel is awake. I say hi and walk to the window sill. Daniel follows me. He looks great in faded twist jeans and a T-shirt that is just the right amount of tight to show off his muscled chest below. His hair is all over the place and his eyes are red – maybe he has been crying, maybe just tired – I feel I should know which, yet I don't. We smoke a cigarette together while watching the night scenes on the street below.

Chapter 15

I slightly lift my head, close my eyes and make the sign of the cross while standing about twenty metres from the entrance to Little France Hospital. A prayer goes through my mind, though "prayer" is not an appropriate word for my thoughts. I am not religious, but I am desperate, and I will praise anyone if He or She will grant me just *one* measly successful experiment. I am spending hours and hours on the bench but apart from this lucky *Angewandte Chemie* paper, which I still don't understand after reading it three times, this is not a fruitful investment of time and energy; I face only *failure* after *failure* after *failure*… I can unearth not one single result that might lead to something promising. I read scientific articles like an automaton, and have intellectually grown from barely knowing how to spell cystic fibrosis to becoming some sort of expert. But as long as I remain unpublished in the field I might as well play Sudoku all day or amble along Berwick beach with a metal detector or otherwise live into the solitary existence which will be my destiny. No one will take me seriously. I won't be selected to give a talk. I won't win a single prize. And it might get damn hard to finish my PhD in time.

I have had a funny feeling in my stomach since I woke up this morning. I feel adrenaline pumping through my body. I am apprehensive, nervous, *very* nervous indeed. I might finally be on to something. I better be, because I did not inform Mark about this experiment and it devoured almost all the valuable "Raetz antibiotic" which we could never re-stock. He won't be pleased, not in the slightest, to hear that I wasted all this precious material on an experiment that is not his baby. *I will hang if it didn't work.* But I needed to do it, out of self-protection, to preserve my sanity, my hopes, my

© Springer International Publishing AG 2017
K. Bodewits, *You Must Be Very Intelligent*,
DOI 10.1007/978-3-319-59321-0_15

dreams. I had to pursue something I believed to be more promising than Mark's megalomaniac, pie-in-the-sky pseudo-projects.

I know as soon as I enter the lab and open the 37 degrees stove to take the 140 agar plates with bugs out, I will have the answer. I take another deep breath before opening the glass door of the lab. I felt my back hurt on my way to university, begat by standing in an awkward position for a full day followed by a night of insomnia. However, right now I feel nothing. Something inside me wants to run away, get on a flight to Bulgaria, any place where I can avoid this terrible risk of abysmal disappointment. But curiosity drives researchers. I have to know the result. Not knowing the answer would eat at me for the rest of my life. My nerves can contrive escapist fantasies, but I know there is no option: I must open the stove. One more quick prayer, a yearn for the promised paradise (maybe Jehovah's are right, who knows?) and here we go…

I lift the handle up to open the metal door, oh-so slowly… *My plates are still there, that's a start.* I take a few out and look at the growth patterns. *It looks good (gulp).* Nervously I spread all 140 plates out, over two lab benches, knowing that if Brian comes in he will bawl at me for invading his work space. I don't care. I don't care about anything good or bad, I don't care if the world is going to end tomorrow, right now all I care about – in the entire universe – is the result. I look at all the plates and compare them with my predictive notes… *correct… correct… correct… correct….*

Tears of happiness are welling up in my eyes. I want to shout, scream, and throw myself to my knees on the floor. I can hardly contain it. But I do. I keep on whispering, "Got it, got it, got it… I FUCKING GOT IT!"

A familiar groaning noise startles me out of my loving gaze at the Petri dishes (and it is love, no less). "Hi Brian," I say, way too happy before turning round to face him.

The expression on his face softens as soon as he meets my eyes. It is almost as if he could be a nice person. *Miracles happen.*

"You got some good results?" he asks, looking at his plate-covered bench.

"Yes," I say, starting to stack everything on piles on the bench I share with Hanna.

"What is it?" he asks. *Since when are you interested in my work?*

Nervously I explain what I did.

He lifts up some plates for closer inspection.

"That's pretty cool," he says. *Is he now even smiling, sort of? Does Brian smile?*

"Yeah. I think so, too."

Without words, he makes it plain that our conversation is now over and I should get the plates off his bench pronto. *Once an arsehole…*

I hastily slide the piles of plates onto the corner of my bench before Brian comes up with the idea to bin them and ready myself to leave. Just before closing the door behind me, I say: "Brian?"

"Yes Karin," he replies, irritated now.

"Did Celtic play?"

"No, why?"

"Just curious!" *I need to tell Mark!*

I rush out of the lab, almost bumping into poor Sharon who has her left arm in a sling today.

"Broken arm?"

"No, dislocated elbow. Will be fine again in a few weeks."

"Rugby?"

"Yeah, tough game on Saturday." *Crazy girl.*

"Get better soon!"

"Thanks."

I cycle back to town, away from the fields that had gotten bright green after a wet May, and away from the tidy and sterile work environment. Happily I enter the School of Chemistry and rush to the lab.

"Mark has been looking for you! He did not seem pleased you weren't here yesterday and this morning," Quinn says with a smile on his face.

"Where did you tell him I'd gone?"

"I told him you probably went shopping." *Arsehole.*

"I might wring your neck one day, but I guess you don't mind."

"That would mean I escape my PhD, yip, I'm fine with it."

"I will do a half-baked job so you will survive."

Quinn laughs. "Don't worry too much about it. It was just a joke and Mark wasn't too bothered. Plus, Hanna kind of saved your ass by telling him you were probably in the hospital."

"Which I was!"

"Yeah, right." *You are such a beefy moron, Quinn.*

I drop off my coat, walk straight to Mark's office and knock on his door. He instantly calls me in and frees a chair for me to sit on.

"Where have you been hanging out?" he asks.

Even though his voice sounds friendly, I feel my hands trembling. *He probably thinks I have indeed been shopping.*

"At the Royal Infirmary. I've got some results I'd like to show you."

"You got the plasmids ready?" he asks, critically. *Stop asking about those bloody plasmids! We both know I don't have to go to the hospital for that, do I?*

For a short moment I feel deeply disheartened, and consider stopping the conversation here and now; slamming his office door shut never to return. But I know it's just another demented fantasy – they occur increasingly.

"Well... no. I am totally on it, and think I might almost have them... but I would like to talk about something different. From home, with my laptop I analysed over twenty genome sequences of all the available *Burkholderia* strains and predicted, based on the presence or absence of one particular gene, which type of superbug would show resistance to the antibiotic we got from Prof. Raetz and which would be killed. Yesterday I went to the hospital to actually prove my hypothesis with some self-made disks.... and it worked." *Please don't ask how much antibiotic is left.*

I show him all my notes and repeat a second time exactly what I have done, and how the results support my hypothesis. Mark inspects them carefully before he finally says: "Fantastic! We need more data, but that – together with the plasmids – might lead to a nice paper. Well done! So you see... you fail, and fail, and fail, and eventually you have success. You're doing a great job here, Karin. Let's send all the data to James, he'll love it. I'll ask him straight away to pay a flight to Venice for you." *Awesome!*

"Venice? To the *Burkholderia* conference?"

"Yes, it's in two weeks. It's too late for you to give a presentation, but you can join us. Hanna and Brian will both give a talk." *Porchetta, smoked pancetta, Rosso di Montalcino, ore di sole... At least I learned something from my posh in-laws!...*

"I would learn a lot," I say.

"Okay, I'll get an email ready for James and cc you in."

I stand up to go back to the lab. "Oh, by the way Karin," Mark half-shouts now as if I were not standing just one step away from him but already down the corridor. I turn to him. "Yes Mark?"

"I've been wondering if you could redo the marking of this practical course you have been teaching," he says, handing me a pile of papers containing the exercises the students had filled out during the first year chemistry class I taught. "You can't let them fail," he adds. *Separating the wheat from the chaff just to bag them together anyway. Sisyphus would appreciate our pointless exercise in Bully-The-PhD-Stooge...*

"I only let the students fail that either didn't show up, were extremely demotivated and didn't even bother to answer the questions, or really did not have a clue what they were doing."

"I know, Karin. But you can't let them fail. You have to make sure they show up and give you all the answers."

"I beg your pardon? I guess that is the responsibility of the student?"

"It used to be."

"I'm not a babysitter, Mark!" I say, upset now.

„ No worries. You will get your degree. Just swallow..."

"Sorry, I can't do anything about it," he says lifting his shoulders and pressing his lips together. "Make them pass," adds he. *Oh Jesus, you must be kidding me, but I know you mean it because you are long past caring about truth and moved in a political world where integrity is just a handicap.*

Chapter 16

With two small bumps the plane makes friends with the ground. We taxi to a quick halt in front of the airport building which is really just an outsized shed. This is not Amsterdam or Paris, and even little Edinburgh Airport is more imposing than the grandly titled "Aeroporto Marco Polo Di Venezia." It's desolate outside the cabin window; just a few men wearing orange safety jackets driving luggage carts through sheets of rain while the windsock indicates a near-gale. The pilot probably announced the weather before landing but I didn't understand a single word. I'm not sure whether this was due to his Italian accent or because my eardrums had sort of shut down after three hours of Mark talking at me, pinned like a lab rat in a metal tube eight kilometers above sea level. My spirits are low. This might be Italy and it might be late spring, and we might supposedly be in the exclusive, high powered world of international scientific confabbing... but leisurely notions of sipping Soave on the hotel balcony in the evening warmth have evaporated in boredom and rain.

Hectically, Mark starts to stuff the preposterous amount of papers which he had taken out to read back into his bag. I doubt there is one single person on Earth who can read so much in so little time. In Edinburgh it had been – for a very short moment – reassuring that he had brought all this reading matter; certainly no time to chat, I thought. Optimistically I had gotten out my headphones, but he started his monologue before I could plug them in. At first, he talked a bit about my project and latest results. However, with great enthusiasm he very quickly moved on to Babette's work with cofactor-dependent enzymes which is apparently quite dazzling, though he never explained why. Maybe it is because it is the only project

© Springer International Publishing AG 2017
K. Bodewits, *You Must Be Very Intelligent*,
DOI 10.1007/978-3-319-59321-0_16

with which he has experienced some (very minor) successes in the last couple of years. After all, it was *this* project that recently convinced some funding body to donate a handful of pounds to the research pot and the finances to hire Barry; his new academic star with remarkably well-hidden talents. Mark showed me 3D structures of proteins Babette is working on and, about an hour into his bragging about how good she is, I got bored enough to risk a comment, "You do know that Babette is not the easiest to work with?"

"Yes," he said in a tone that suggested this was as important to him as a light breeze in the summer of 1452.

"She doesn't talk to anyone unless we have a lab meeting. She stamps through the lab like an elephant. She slams doors. She groans loudly, wheezing and making other noises that stop anyone from concentrating," I added, in the hope that it might yet trouble him.

He just nodded as much to say, 'receipted and filed and forgotten already.' He immediately flicked to the third page of another paper and started to talk about how the cofactor fits into the enzyme pocket.

"Have you read this paper?" he asked, after a lengthy explanation.

"No, I'm working on a different project."

He shook his head. "You should read this!"

The tone of his voice was unfriendly, harsh. However, he actually corrected himself and abruptly reverted to a friendly tone, showing that he was aware we were just adults on a plane chatting publicly. Alas, he didn't read on the journey and he didn't change the topic of conversation, so it was an awful flight.

Both of us had been unhappy at check-in when we realised that we would be sitting next to each other during the flight. But what can you do if you find out that the cute flight attendant behind the desk gave you the wrong seat? For both of us, at least for my poor eardrums, it would have been better if he had simply said, "Karin, I have no desire to sit next to you whatsoever, can you please swap with James?" He could have even wrapped it in something nice like, "I do not want to be rude, but I have to discuss some things with James. Would you mind…" I wouldn't have minded one bit… However, for me to say something like that to him – the man who asserts his authority with his every utterance – would have been signing my own death-sentence; "Mark, would you mind swapping seats with Hanna because I think your one-sided, self-important monologues which I am supposed to regard as engaging conversations might actually make me weep…" Of course I would have worded it politely but deep down he would know that is what I meant. At some level Mark must be aware of the world's low estimation of him – it's a fact he seems to fight hard to keep out of his reality. My hopes had rested on James to save the day but I suspect James was secretly delighted to sit four rows behind us.

The seat belt sign is still on but Mark's body language strongly suggests he wants me to stand up so he can do the same.

"I find it funny to watch all those people standing up in an airplane. All those bodies pressed into a somewhat bended state due to the slightly tilted seat in front of them and their heads getting pressed awkwardly against the low luggage compartments. They look like crippled giraffes with broken necks. It's pathetic. And for what? It doesn't go any faster..." I say, pretending that I did not notice his urgency.

Mark laughs a bit too loudly, and actively presses his arse cheeks into the seat again. He is inferring he never wanted to get up but we both know better. For one fleeting moment I feel empowered over Mark. *I* made *him* feel uncomfortable – wow. It's a curious little moment, unique in our relationship thus far.

At passport control Mark strides over to James and Brian. Hanna gives me a friendly smile and whispers, "I felt sooo sorry for you all the way..."

"I felt sorry for myself. But hey, I got a three hour lecture about cofactor-dependent enzymes and have a new reading list..."

We reconvene at the exit and together rush through the pouring rain to the shelter next to the taxi stand. A short, slightly fat man looking like a clot crumpled into a striped shirt drops his cigarette on the pavement. With cheap leather-look slip-on shoes he taps on the smoldering cigarette and opens the door of the little taxi bus for us. When we are all in, James hands him a piece of paper and Clotty signals that he understands where we need to go and without any words being exchanged, we depart the airport.

Hanna and I sit on the rear bench-seat of the little taxi bus while James, Mark and Brian are on the seat in front of us. For the first ten minutes it is quiet. Mark and James exchange a few random sentences about the weather, their previous Italy experiences, Sharon's injuries and football. James couldn't care less about football but for Mark that doesn't matter. The first few meetings I had sat with both James and Mark, I thought I was too stupid to follow their conversations, but then I realised that their speeches run parallel and mostly don't connect with each other. They both broadcast their own stories, they just happen to be in the same room.

It is all so innocuous but then Mark suddenly turns round to Hanna and me. Out of the blue, he issues an anxious and utterly bizarre warning: "Don't talk to anyone about your research!"

James slightly shifts his upper body, also towards Hanna and me. He turns his head, with some difficulty, to look over the back of the seat to face us. "Be careful. Do not share what we work on. Your data isn't safe here."

From the expression on his face I think I read that he has difficulty with this message; he looks sad somehow. Brian nods along without looking at us. He always agrees with Mark and James, his factory-set reaction to superiors of any kind.

"Alright," Hanna and I mumble simultaneously.

I feel confused. Were we not, as I understood it, four generations of scientists – PhDs, postdoc, lecturer and professor – attending a "scientific working group" representing a university renowned for its scientific excellence? Were we not travelling to Italy to meet researchers from all over the world working on the same superbug? A bug that uses penicillin as a carbon source, a bug that forces paediatricians to separate siblings, a bug that kills; a bug that matters somewhat more than petty academic rivalries – no? Were we not going to this meeting precisely to share our know-how? To discuss working towards a treatment together? Apparently not... Those few words with the Italian radio in the background altered my simple perceptions entirely; it now seemed we aren't far from bit players in a Godfather movie driving into the lair of the enemy...

We head up the driveway of a hotel that looks like a lonely outpost – sort of plonked in the middle of a field. Apart from the torrent falling from the sky, there is no water in this part of Venice. There is also a distinct absence of Gothic arches, Byzantine domes, Baroque grandeur, Renaissance windows and indeed anything to do with Venice. There is however, wispy grass aplenty. *They really oversold the conference: "Scientific working group in Venice?" This is not Venice! This is the neighbouring town of Marco Polo! It is like promising someone London and sending them to Dartford.* Clotty opens the doors for us and starts to pull the suitcases out of the trunk. His arms are just so too short to make it look effortless. It's the only entertaining bit in all this weirdness.

"You have time to check-in and refresh yourselves; a bus will take us to the conference hall in an hour," Mark states.

I walk the stairs to the second floor and let my body flop onto the large double bed with all my clothes still on. I consider my options for possibly having a nap, while marvelling at the absence of interior design. I had got up at 4:00 a.m. this morning to be at the airport on time, after a full week of long days and evenings in the lab. I had desperately, and unsuccessfully, tried to get a few more results before heading to Venice – to have something to discuss with the experts in my field. I am dead tired but the chances of recharging are grim. I could sleep for half an hour, but that would be just the amount to make me desperate to sleep for hours. I better get ready and see if I can get my hands on a strong Italian espresso or three.

The conference starts early afternoon. The room in which we will spend our next three days looks more like a large winter garden than a lecture theatre. We are about seventy scientists, working in different countries at different universities, sharing the same topic for years. Many seem to know each other. Most, if not all, seem to know James. He has guru status in the field, having worked on cystic fibrosis and its associated bugs throughout his entire career. And he is a curiously strong, low-key presence. He always wears a smooth shirt with the upper three buttons open; fluffy grey chest hair jumps from his shirt and is partly flattened by a shark tooth on a leather band around his neck. The surfer-style necklace suggests youthfulness but the years are catching up with his face. His eyes are always wide open, indicating enormous engagement with his work, but they have grown watery with time, and the iris colour has almost faded to grey. A few toothpick-thick hairs sprout from his nostrils and his face is wrinkled. Yet it seems like every wrinkle tells its own story, as if his life and research is engraved in them.

James relishes telling stories from research in the past, the history of the disease and about epidemic outbreaks that killed large numbers of cystic fibrosis patients. Every time Mark and I visit him in the Royal Infirmary he tells his stories. He sits on a large office chair, surrounded by lots of old woody plants that have taken many years to nurture and are strewn amid bewildering piles of papers and books. A replica of an early microscope sits atop a shelf above his head, while he shares what they knew about superbugs forty years ago compared to what we know now. If he could talk about Scottish history and his own childhood in the same way, I would love to be his granddaughter sitting on his lap listening to him for hours.

Today I am proud to be connected to him, to be at this conference which he will open with a keynote lecture. He stands in front of the audience, waiting for a young guy to transfer his PowerPoint slides from USB to computer. He is waiting patiently, not showing any nerves. When all is ready he starts talking quietly, but with that same passion and emotion I have heard in his office. For one hour he is on home ground, talking about times gone by and the scientific developments he has experienced.

He expresses deep care for the patients he has worked with. People hang on his lips, including me.

A person that made a life choice, who has a single-minded research focus. He is married to science, married to academia. He will keep his office, with the microscope and woody plants until death or disease separates him from them. Could I ever be passionate about something the way he is? Could I ever devote all my time and the expenditure needed to become such a refined, admirable researcher?

James finishes his talk with the words that he is hoping for a productive conference with lively scientific discussions bringing us closer to the target of developing drugs and getting the superbugs under control. *So we are part of the same* familia *after all?* He receives applause and much shorter presentations by other researchers follow.

They all present work which has already been published in scientific journals – I have read it, and I am tired. A short discussion follows every presentation, but nobody in the audience seems to give a different perspective on the research which has been presented and nobody offers their thoughts on how to continue the work. During the fourth presentation I whisper to Hanna: "They all just present stuff that has been published already."

"Yeah," she says, looking as bored as I feel.

"What's the point of that?"

Before Hanna can answer my question the young guy next to us, who neither me nor Hanna knows, interjects, "In settings like this the risk is too high – you cannot present something that has not been published yet."

"What risk?" I ask.

"The competition is fierce in this room. We all work on the same topic. It is too easy to steal each other's results."

"But are we not supposed to work together instead of against each other?"

He chortles at my lingering naivety and whispers: "In an ideal world yes, but that is not the reality. The reality is that we all want and *need* to be the first one to discover things."

I look at the young guy who sports half-long hair, a three-day beard and a light pink shirt. He is a few years older than me, maybe a postdoc or a young group leader.

"Are we not meant to share scientific advances, to improve the lot of humanity?"

I'm playing devil's advocate by now and he laughs openly. "The first priority is to secure research funding. To get the best funds, you need to be the best… My PhD supervisor always told me: In academia there is no second place. You either win or you lose."

„ In academia there is no second place.
you either win or you loose. "

"So, what exactly are we all doing here then?"

I'm starting to feel slightly rude for talking through someone else's talk but it is fun coming across this blunt fellow. He lifts his shoulders in a way that clearly states he doesn't really know the answer to my question. After a thoughtful break, he gets back to me: "We have a beer together and, on some projects, we do collaborate with each other".

Okay, so we are just pretending.

When the coffee break is announced, he sticks out his hand to me. I shake it briefly. "I am Marco Julienne by the way."

"Ah, you work for Professor Wittburg in Vancouver? I am Karin, working in Edinburgh. We had email contact before, about the tri-parental mating technique. Remember?"

"Of course I remember!"

"What a coincidence! How is your research going? Did you get the bugs manipulated?"

"Yeah, it's fine, it worked."

"Are you guys following up on the WaaA project?"

He nods. Slightly uncomfortably, he looks out of the window, as if he is keen to stop the conversation.

"Did you guys manage to purify it?"

He turns his head towards me. His eyes move up and down while he exhales through his nose.

"I am sorry Karin, I would love to speak with you, but I am not allowed to talk about my research."

After the conference dinner I collapse on the bedcovers and find myself gazing at black mouldy circles on the ceiling, which seem to symbolise this tainted experience for me. I am unable – or desperately unwilling – to believe that this conference has nothing to do with a group of bright experts from all over the world trying to solve a problem. Yet we are people keeping our data secret from each other until we publish it. We are people who are scared of the dishonesty and bad behaviour of others, and of course suspicious people are the ones to suspect... We are all fighting our own corners, we see each other as opponents, we are looking for selfish affirmation while pretending to search for scientific truth. We are all desperately yearning for recognition – after long and lonely hours on the bench. We are people with a professional – and financial – existence dependent upon being first, on being the winner. It's all a mad hoax, our research has nothing to do with ideals...

Wearily I reach for my blinking telephone on the night stand. The first message is from my mum, who has clearly forgotten that I am in Venice and just phoned for chit-chat. "Hi Ka... just curious how you're doing... all fine here... dad has man flu, thinks he'll die, just a cough, nothing serious... give me a ring when you have time, no urgency..."

The second message is from Daniel, and his voice disturbs me; "Hi Ka, it's me. Listen, can you call me back? I'm sure you're busy, but there's something with Karel. He's in hospital. Just call me back, will you? Miss you, bye."

Karel, a friend from my undergraduate days, is not known for his healthy and safe lifestyle choices. I take the shower I have been craving, and can't help worrying about him. I wrap my body in the largest white towel available, which is not large and only white-ish, and flop back onto the bed to phone Daniel.

"Hi, it's me. What's with Karel? Bike crash again?"

"No, Ka. No," says Daniel in a completely unfamiliar tone.

"So… what…" I hear my voice sounding scared.

"Ka… I'm going to be frank. Karel has got cancer and he won't survive long."

"What?!" I squeak.

"Apparently he had a strange pimple on his neck that was getting bigger by the day and his PhD colleagues told him to go and see a doctor, which he did. That was a couple of days ago. It transpires he has a *very* aggressive, deadly form of connective tissue cancer. They can't do anything for him."

I feel a pressure behind my eyes. I try to swallow but my throat won't clear to let me speak. I take a few deep breaths and try to phrase something but all that comes out is a shrill noise. Eventually I manage to talk in a horribly high-pitched voice, "But they can't *know*? Right? Not so quick."

"Apparently they can."

"But… How long?… What's his life expectancy?"

"They can't tell exactly, but probably a couple of weeks."

"A COUPLE OF WEEKS!!!… Fuck… he can't finish his PhD!" I say quietly.

"That is probably not his biggest problem."

"True… Of course…"

We stay on the phone for at least half an hour, though with long breaks in the conversation. Daniel tells me when he heard about it and explains that Karel unsuccessfully tried to reach me this afternoon.

"It is probably too late to phone back," I say.

"Yes, he is really tired, sleeps most of the time."

"When are you going home to see him?"

That question cleared my mind. *To hell with all my experiments… Mark wants results, and I still have no plasmids for the US and the purification of the protein I just expressed before Italy, and…* As if Daniel is reading my mind on the other side of the line he goes on, pointedly. "You are going home, right? To visit Karel? You can use your work as an excuse not to see me, but you are not seriously going to say that you are too busy to visit a friend who is suddenly dying."

"I didn't say I wasn't going. I just need to figure out when would be the best time."

"Yeah, right. I am sure you will make a good decision."

Daniel is deploying a condescending tone which triggers me at the best of times. It is the same tone he uses when I snack before a late dinner, fretting that I will lose my appetite.

"Why the patronising tone? I'll be back in Edinburgh in less than 36 hours. I will book a flight to the Netherlands immediately. Meanwhile do me a favour and keep your boiled-up resentments to yourself!"

I press the red button to hang up.

I throw the towel on the floor, turn the pillow round to have the dry side and pull the blanket up to my chin. I am angry with Daniel, but somewhere inside I know what really infuriates me is that Daniel read my mind well. *Would I prioritise the progress of my PhD over the last chance to see a friend alive? Am I as mad as this mad game we are all playing at this conference where we're all determined not to confer?... No! I'm telling Mark tomorrow I'm going home this weekend.*

Chapter 17

I sit down on the window sill resting my legs on the rocking chair in front of it. A bottle of wine stands next to me, my hands unwrap the cellophane of the packet of cigarettes I bought on the way home – it's an exciting, urgent little ritual for us sad addicts – and Tom Waits' smoky voice is sound-tracking the moment perfectly. It is only Thursday and I feel emotionally, physically and mentally exhausted. Mark had informed us during one of our ill-fated coffee breaks on Monday that he expected all of us to attend a scientific conference lasting all of the next day, in Dundee. So today had been a predictable mixture of boredom and bad food in a concertinaed lab week; I had worked long hours to stay on track, despite my schedule being somewhat pointless.

Yesterday I acquired an MSc student to supervise in the lab. He is clearly a passionate researcher; quite tirelessly devoted to discovering which of our female bachelor students might be willing to slap the mattress with him. Life beyond that obvious purpose seems to consist entirely of annoying me. He is witty enough, but useless; doesn't know how to hold a pipette and can't make serial dilutions or any other calculation a ten year old could be taught. In keeping with the general absurdity of my PhD, the *Angewandte Chemie* paper and the latest results I gathered have boosted my motivation. I have the feeling that I could make a footprint in the scientific community, but a student screwing up my days is not helpful.

I let the side of my head rest against the window, inhale deeply from the cigarette and watch the pedestrians, cars and buses weaving their way along Gorgie Road. It's busy even though it is after 9:00 p.m. A guy strolls past; he's about my age and, based on his looks and gait, I momentarily believe I am looking at Karel. But of course I am not a necromancer and Karel

© Springer International Publishing AG 2017
K. Bodewits, *You Must Be Very Intelligent*,
DOI 10.1007/978-3-319-59321-0_17

hopefully has better things to do in his after-life than haunt Edinburgh eating saggy chips from a newspaper.

After Venice, I had travelled back to the Netherlands to visit him in the hospital. It had been a sunny day and we sat ourselves on a bench in front of the University Medical Centre. He wasn't allowed to leave the premises – too sick. The pimple on his neck had been removed but was making a very visible comeback. More cockle-like growths were appearing on his face. Though the cancer was so clear and present, and evidently uncontrollable, Karel dreamt about his future; beyond his PhD, far away from the prick who employed him. I guess he was hoping a miracle would happen and he would survive. I found it difficult when he talked of this future which would never be, but of course I just listened.

Though I respected his optimism, I didn't want to boost his ill-founded hope. At a deep level he knew as well as anyone there was no hope. Through his optimistic words you could sense his pain and sadness, and that he was scared. Karel, too, had dreamt of becoming a doctor and making an impact on the scientific world. He too had devoted his life to science, and he was good at it. But fate curtailed any meaningful outcome. Despite Karel occupying a bed in the same hospital where he was doing his PhD research, his supervisor never visited him in his final weeks.

A couple of weeks later I travelled again, this time for the funeral. Karel's PhD supervisor offered the research project to Karel's best friend the day after he died – even before he was buried. "The project has been stalled for too long already," he had said.

Karel had been sick for merely a few weeks. Even so, apparently the research was lingering badly. I never met his boss but I wondered what sort of half-man he was?

Are these the people that will make an impact on the world? Does academia really let people through because they are good researchers, regardless of their humanity? Is it really only good publications that matter? Are these the people we want to have as role models for future generations of PhD students? For our children? I like to believe that only good leaders can build a winning team, but evidence is sometimes scant…

During the days immediately following Karel's death I tried to understand what Karel must have felt and thought when he heard he would soon be dying. Then my mind blocked the story out, completely, as if it never happened. Tonight Karel has re-entered my mind with surprising force. Did he regret spending his last two years in that lab? How did he feel knowing that he would never finish his PhD? Did it matter to him at all? Is it a superficiality or did it appear so to him? I suspect it did, but I don't really know what was going through his mind. I don't understand it, and somehow I hope I never will. My face is pressed against the window, I down a glass of wine and light another cigarette to keep my emotions under control. Despite my efforts, I feel immensely sad and then a beep from my laptop punctures the mood. Once upon a youthful day I tingled to the sound, thinking somebody was trying to reach me, probably just to say hello or send some silly message but, even so, someone wanted to share their world with me. When I was in China I was Pavlov's dog racing to the call of the beep to hear from my precious friends back home. No more. Today that sound triggers anxiety.

My heart is lurching. *It is him. It must be him.* Mark has crept right into my life like an omnipresent parasite gnawing away at my well-being. Day and night, my life revolves around my PhD now – or, more accurately, around Mark. Between 9:30 a.m. and stupid o'clock in the evening, there is always a chance that an email is being pinged my way, quite a high chance. I doubt he has the faintest idea what they do to my heart rate, neck muscles and breathing.

Slowly I walk to the laptop and my hands are shaking when I open my email. *Praise the Lord – or whoever – it is not Mark. It's the girl in the US, who I only know through email.*

She is in Prof. Raetz's research team and is tasked with working with my plasmids. Somehow, I still can't believe that this world-leading research group is willing to work with us. I can't help suspecting they have rumbled us and put a loser on the case; like the "researcher" who is asked to carve the Halloween pumpkin, or fill out the health and safety forms, *or* work with us… However, this is just weary, self-loathing prejudice; so far nothing suggests she is incompetent. I open the email:

Dear Dr. McLean,

Dear Karin,

The plasmids you have sent over to the US are full of mutations (see the sequencing data attached). Could you please send us the right plasmids without mutations. This way, we will not be able to continue the project.

Best,

Wang

I'm reading the email a second time and after the first sentence I feel the stress sending my body into meltdown. *W.T.F. have I done? Oh no, this is the worst thing I could have done…*

If Mark has already built up credibility in Raetz's lab I have definitely destroyed it utterly, by sending plasmids that are just rubbish. My whole body is shaking now and I feel deathly cold. Frantically I walk around the living room without purpose. *This isn't happening! Please tell me this isn't happening. I will just wake up in my bed tomorrow and everything will be fine. It's just a bad dream, nothing else, maybe I'm hallucinating, I've had precious little sleep for days...*

Eventually I open all the attached files. It doesn't take long to see that Wang is correct; the stuff I sent is indeed full of mutations. Just at the moment I grab the pack of cigarettes on the window sill with shaking hands, my cell phone rings. Mark rarely phones me, he prefers email. But this is too serious. *He will verbally destroy me.* Like a quivering wretch who cannot face her obvious doom, I walk to the coat in the corridor and take the phone out of the pocket. On the display I see – *Hallelujah* – it isn't Mark, but Daniel.

"Hi," I say.

"Where are you?"

"At home."

"But you were going to join us for a drink tonight."

Daniel sounds upset. We haven't seen each other much in recent weeks. Most evenings I come home when Daniel is peacefully snoring. We see each other in the morning sometimes, but it's too brief for in-depth conversation about anything. Our relationship never thrived during sunrise. When the love motor was still going we hit our heights in the evenings.

"Drinks?" *Oh fuck yes, drinks.... Ah damn...* "I forgot about it."

"Ka! You promised me you would come. These are my goodbye drinks from the lab!"

I'm quiet for a moment. I did promise him I'd come to the drinking session, which is finally taking place months after Daniel left the lab. The professor he worked for takes things nice and slow. He got his remarkably comfy chair by reaching the end point of the university career programme, which Mark is now pursuing. He hasn't published for over a decade. I doubt he even noticed that Daniel's lab book remained astonishingly bereft of writing till the day he departed. Prof. Comfy-chair seems to be happily snoring away in his safe position. I dare say if he puts his mind to it Daniel could snore away the years too.

He has been unemployed for months now. I suspect he mostly dreams the days away, while sitting at the desk in the living room as if working. He hasn't left the flat much since he finished at the university, and is mostly home alone. Apparently it had given him time to think and recognise that our relationship is dying. I would argue it is already dead, but Daniel still sees it more positively. "We just need to spark it all up a bit. Get involved more in each other's lives…" he had said, desperately, on Tuesday, after one of our shouting matches.

He was sat at the kitchen table using the empty plate in front of him as an ashtray, when I arrived home just before midnight. "I thought you would be home for dinner," he had said.

I couldn't tell if he was just really annoyed or about to cry. "That's what I had hoped as well," I said neutrally, while staring at the second plate which was still full with pasta carbonara.

It would be true to say that anything not directly related to my research or other obligatory PhD duties has long ago faded into insignificance for me. I could not care less about being home for dinner or not.

"I told you I would cook something and you didn't even bother phoning to tell me you wouldn't show up."

He had almost yelled at me. His face had turned red, his neck and ears followed. With some guys I find it very sexy if they get angry. It can trigger me in a good way. But Daniel wasn't triggering me. He looked like a beaten dog in the corner, giving one abject bark before retreating back into his kennel.

"I forgot to call."

"You forgot?" he positively screeched. "When was the last time we actually saw each other?"

"This morning," I replied, hating myself for being so literal and superior.

"Fuck off, Ka. We never see each other because of your never-ending experiments!"

"I'm not sure if you're aware of it but I'm doing my PhD, Daniel!"

How tempted I was to add that I have, in contrast to him, finished my MSc long ago and didn't take six months for a single PCR reaction. That I am at least doing *something*.

"This stupid PhD has taken over your entire life!"

"I know. Maybe a PhD is supposed to take over your entire life?"

"No! It's insane, Ka. Not all PhD students spend their whole life in the lab."

"Maybe they have plenty of results and I don't? And believe me, I'm not the only one in the department in the evening."

"Don't tell me that all PhD students bring a library worth of papers to read over Christmas holidays, Ka! And that they think twice before visiting a dying fr…"

"Don't even go there!"

I had regretted my moment of hesitation on the phone in Venice, I had horrified myself instantly, I was in shock I think. Daniel hadn't brought it up again – until that evening.

"You promised me you would put in an effort," Daniel says sadly, on the other side of the line. *Did I?*

He adds, "Our relationship needs a spark, but you don't do anything for it!" *A spark?*

I think ten thousand volts might be insufficient, but I don't say that. Our lives, which had been compatible in our undergraduate days, had diverged since the start of my PhD or maybe since the end of my masters. It was hard to separate the whys and wherefores but oh-so easy to play the blame game, for both of us. It was sad, in the old-fashioned sense. And of course my stress levels helped nothing. I had promised to put an effort into getting back on track with him. Drinks tonight – that was the first planned step in that direction. And I had forgotten about it. Maybe I didn't really care enough to remember. It had been a tough week in the lab and, in the way of these things, I do not know if that signified a lot or not a jot…

"I'm sorry, but this is not the right time for me to give a shit," I reply sadly.

"Are you okay?" Daniel asks, in a compassionate tone.

"Nope! I screwed up, made a major mistake."

I tell him about the email, about the Raetz lab and how gutted Mark will be. "What are you worrying about? Just send them the right tube."

"There probably is no right tube, Daniel. I believe I might have forgotten to check the sequencing data altogether. Mark is not going to like this!"

"Oh, your supervisor has other stuff on his mind, Ka."

"No he doesn't."

As I say this, I realise that Daniel still doesn't grasp what Mark is like; that my boss is an unpredictable, short-tempered and utterly inept helicopter supervisor. He crossed a line one day and shouted at me. Since then he has worked up to barking at me freely, regularly, with no restraint. He is not unlike a parent who dirties his hands by beating his child and soon he "has hands so steeped in blood it's easier to go on as go back," as Macbeth sort of said. With every new baby the parent might resolve not to fall into the same habit but, like the lab chief with a new underling, the tempting day duly arrives… Bad habits die hard.

I used to share stories with Daniel but I haven't shared much lately. I've left him in the dark about Lab 262. And this was not the right time to clue him in, to share how worried and stressed I have felt since the start of my PhD – not tonight.

"Are you still coming for a drink?" *Oh God, no. I prefer to drink alone.*

"It's already late, Daniel. By the time I could get there the pubs will be closing."

"Okay."

He sounds sad and hangs up without saying bye.

"Dammit! Dammit! Dammit!" I want to throw my laptop against the wall, ideally against Daniel's stupid picture of a sailing boat. Everything is just bearing down on me and I feel like a failure every which way. There is an alternative plan to dealing with all this; the simpler, and probably much nicer one, of disappearing without a clarifying word to anybody. I could go to Alaska, naively playing *Into the Wild... Continued*. Or anywhere else really, just as long as it would be remote enough that no-one would come up with the idiotic idea to follow me.

But I know I'm going nowhere far, or near for that matter, or in my career either... if I don't buckle down like a good girl – like Pavlov's dog – and look for the source of the mistake before Mark contacts me. On my laptop I search for the link I must have received from the sequencing service in Edinburgh before sending the sample to the US. I open email after email but can't find the right link. Via the link in an email from months earlier I enter the network in which the data must be stored. I scroll through two weeks of sequencing data which the lab produced. There are at least three hundred samples listed but none of them are mine. *I really forgot to bring it and check it. How was this possible? After so many months of trying to create the plasmid, I had forgotten the last and most important step. Where is my mind?!*

I pour another glass of wine. Alcohol might not solve any problems but it can shrink them quite well. I hate myself for cocking up and I hate myself for the happiness I had felt when I finally had a plasmid worth more than zip. I hate myself for being able to continue my PhD while Karel had died. I hate myself for not going to Karel's PhD supervisor to punch him in the face. And I hate myself for hurting someone with bad-tempered, selfish words. I hate myself. I feel tears welling up in my eyes and just let them spread mascara over my cheeks. I drink. I smoke. I listen to the music.

My head aches and eventually I lie down on the bed. Mark has caused me some worrying hours but this is the first time in many months that I lie in bed with my eyes wide open, completely unable to sleep. Over and over again I excuse myself to Karel for blocking him out, and I rephrase explanations for Mark.

There was no way round it; I had to tell Mark the truth. I cocked up. I cocked up big time. I try to push it out of my mind by thinking about other things but it doesn't work. For hours I lie there in torment, still awake when Daniel returns. I don't want to talk so I feign sleep. When he enters the bed he wraps his arms around me. Finally I fall into dread-laden dreams.

When the morning rolls round and the alarm clock fills the bedroom with its terrible sound, it feels like my sleep has been exhausting rather than replenishing. With stiff muscles and make-up everywhere, I seat myself behind my laptop. No email from Mark.

I get myself ready, down a coffee, and cycle to campus, slowly. As per every morning, it's only Babette in the lab, abusing some equipment. She doesn't greet me, and for the first time ever I don't greet her with my friendly "good morning" either. The last couple of weeks I had started to vary the pitch of the words "good" and "morning" and I confess I derived malicious pleasure from the fact that she got progressively more aggressive with the amount of melody infused into my unwanted greeting. She particularly loathed it when I personalised it into a jaunty "good morning Babette." But today I don't care to pour crumbs of salt on her psychopathic wounds. *For what was I doing it anyway? What vile traits has this PhD stirred in me?*

I walk straight to the office. From the corner of my eyes I see Babette lifting her head and looking in my direction. She doesn't look as angry as she normally does, but rather confused. The absence of a greeting foxed her. After a few seconds pause she mumbles something that could resemble "morning". *Did she just greet me? Bizarre. Feels ominous somehow...*

The office comes to life just before 9:00 a.m. When Hanna arrives, spot on at 9:00 a.m., I immediately tell her about the plasmid. I'm not sure why I picked her to tell? Probably because she is working on a similar project and is definitely aware of the importance of Prof. Raetz. Even after the *Angewandte* paper, Hanna and I get along reasonably well, and she will definitely supply some soothing words.

"Don't worry too much about it," she says. "It's shit, but it could have been worse."

"I don't know how it could have been worse."

"Me neither! You hang big time!" Quinn says, sitting next to her with a big smile on his face. *Does your mum know she raised a dick?*

"Don't say that! It took her months to make the plasmid. You know how frustrating it is and how unreasonable Mark gets."

Hanna sounds upset on my behalf.

"Sorry," Quinn says.

Hanna addresses me: "Ask Mark to buy you the plasmid instead. It only costs a couple of hundred pounds and it would mean you're not wasting your time."

It really was a waste of time. I hated myself for spending hours and hours on creating something that could easily be purchased. In fact buying it would have been much cheaper than all the consumables I was wasting to unsuccessfully create the blasted plasmid.

"Forget it," says Quinn, more to Hanna than me. "Wrong research project. If she was working on KBL he would fork out, but she isn't."

We all know Mark is spending most of the available money on a single project and is very unwilling to invest in the others.

"Just try, he might," Hanna says.

At half past nine I see Mark enter the building through the chemical waste entrance below our office. He normally drops off his coat and bag in his office, then visits the lab while his computer is booting up. I wait a few more seconds, trying to psyche myself up for the showdown.

I hold my breath as I knock on his door. He barks me in without even knowing who it is.

"Morning," he says and taps the power button on his computer.

"Hi," I respond feeling a horrible lump in my throat.

"What happened with the plasmids you sent to Raetz's lab?"

His voice is loud, but the tone is friendlier than I anticipated.

"I sent them the wrong ones."

He must sense my nerves. He just smiles, perhaps wishing to calm me down. Perhaps my evidently nervous state has gratified him already.

"That's incredibly stupid. But just send them the right ones."

"That's the problem. There are no right ones. I forgot to sequence them."

"Then you have to make new plasmids."

"I know, but it took me months and months to get a PCR product at all. Making new plasmids sounds easy, but it isn't."

He looks concerned, slightly angry but not as if he might start shouting. I suspect he is worried I might start crying. He doesn't like that kind of emotion in his office. It fazes him.

"What about buying the gene?" I dare to ask.

Mark starts laughing as if this is the funniest thing he has heard in a month of Sundays. "Buying the gene?" he mimics. "No. Just keep on trying!"

I press my lips together. Mark is leaning over his computer and a quick fantasy flashes forward, one in which I smash his head into his computer screen. The tingling noise of his keychain when Mark stands up brings me back to the sad reality – in which he doesn't get his head smashed in. I step over the high stacks of papers on the floor and leave his office.

Chapter 18

With a new undergraduate following closely behind, Barry enters the office precisely five minutes after I sat down on the chair that I can now finally call mine. Bubblegum-Bobline and Diet-Coke-Girl have created space in the office – by leaving, forever. Almost a year after their stipends ran out – i.e. one year out of their lives doing unpaid work – they defended their theses and finished their degrees. *Pffuuh.*

The PhD "ceremonies" were deeply unexciting. They sat their exam behind closed doors in both cases, with only the two examiners; one internal and one external, as is the established practice. After their defence, they came into our office with the examiners to celebrate their success. On each occasion Mark bought a bottle of Brut to celebrate. That there was only one bottle to share between about twelve people rather suggested, to some of us, that our presence was not much desired. But at the same time we were meant to be there, at work. It was awkward, in fact downright embarrassing all round; such a paltry and token homage to an endeavour which took years, a celebration as glamorous as an "employee of the month" bash in the back room of a provincial McDonald's.

I think it was those two occasions, which are supposed to be among the most memorable experiences in someone's life – for some scientists even more important than marriage – that made me realise just how broken and disturbing our lab actually was. Maybe it wasn't Mark's responsibility to organise a nicer ceremony, maybe it was our responsibility? As colleagues

© Springer International Publishing AG 2017
K. Bodewits, *You Must Be Very Intelligent*,
DOI 10.1007/978-3-319-59321-0_18

spending the whole week together, all with the same aim, the same limited funding, and the same lack of work space, perhaps we could have pulled together for a party, but we didn't. Could we not wish joy to each other? Were we a lab of individualists doomed to grudge one another any limelight? Or was it just Bubblegum-Bobline and Diet-Coke-Girl whose negativity soured the atmosphere like an odourless toxin?

Certainly the mood in the lab relaxed when they left. How can intelligent people with common interests and common goals be reduced to this: people imparting happiness by departing, and at no other time. It is not just wrong – it's weird.

Bubblegum-Bobline and Diet-Coke-Girl had hated each other. And everyone seemed to dislike them, including myself. Yet no one ever spoke about their dislike; there was no need, it was so clearly understood, by everyone. As soon as they packed up, and closed the heavy brown door with the number 262 behind them, a silent and yet unmistakable sigh of pleasure filled the room. Their names were crossed out on the telephone list in the office and removed from the door within a week. I don't know who crossed them out, but no one objected. For the rest of my stay in Edinburgh I will never see them again. And Mark will never mention them again.

I found Bubblegum-Bobline and Diet-Coke-Girl equally irritating, but in different ways. Diet-Coke-Girl was too sensitive, too emotional for me; an incredibly unhealthy, high-pitched drama queen, in my eyes. I struggled with her, but Mark seemed to despise her. He couldn't stand having her around. She was very different from Mark, at heart a dour man who squirmed in the presence of normal feelings. Diet-Coke-Girl's emotional incontinence was dreary for all of us, but for Mark it was the stuff of nightmares.

Maybe Bubblegum-Bobline-Girl was Mark's best sparring partner. She ran roughshod over the hierarchical structure and sparred on equal terms. She did not give a damn about him, or about hiding it. I would have admired her if she had at least smiled or said a friendly word to me just once, but apart from the drunken-girlie-chat at Christmas dinner she didn't. It was a mystery whether or not Mark disliked her. He seemed slightly scared of this woman, who rounded off all their interactions with choice words like, "Thanks for your input, Mark! It's an absolutely idiotic plan and I'm not going to do it!"

They weren't bad scientists. From what I heard and later read in their theses they were both good. Bubblegum-Bobline-Girl built the fundaments (and even published) a research topic Mark is still working on today. Part of Diet-Coke-Girls' thesis would easily have been accepted for publication in a peer-reviewed journal, but Mark hated her so much that he never read her thesis before submission. His old boss, Prof. Gilton, had to tend to that. But even if Mark had read it, he would probably have opted out of them writing a paper together. This would not be because of his lack of input – that's so passé it doesn't merit comment – but because he is so bloody-minded he would cut off his own nose to spite his face.

Now I wonder: Were they really such awful personalities or was that just how I experienced them? How would it have been had we met as colleagues in a different lab – possibly rather nice, sadly. Mark badmouths others to me and apparently badmouths me to others. Only Lucy and Logan manage to steer clear of Mark's rants, mostly because he isn't interested in their projects. We all know what is going on, and feel we should be able to rise above it, but in reality we can't. Mark is sitting in our ears like a menacing earwig repeating over and over again how incapable the others are. And even though we know he is a brutal character assassin, our brains cannot help but look for evidence to confirm his incessant put-downs. Consequently we all feel defensive and traitorous. It's a wonder we interact at all – a triumph for the social animal inside us all.

I muse that, no matter what, we have to be happy with the small things in life. Two departures have freed up two desks and two chairs; for Logan and me. Mark had organised a few more desks in a room on the other side of the building; for Babette, Erico and Barry. Perhaps he could not bear looking at Barry's sad face any longer and he hoped that more office space would cheer him up, but it didn't. I guess their new room, inhabited mainly by postdocs of Barry-ish hues, doesn't inspire them. It is not very practical either. They have to cross the whole building to get to the lab, which is disastrous when most experimental waiting steps tend to last five to twenty minutes. But they luxuriate in an indulgence many of us envy. They have desks in a clean office where Mark cannot easily sneak up on them. I am happy enough just having a desk facing the window, next to Lucy.

"Expectations versus reality"

Barry is standing next to me, looking through a chemicals catalogue, with his bored undergrad close by. As soon as he notices that I am doing nothing more intellectually challenging than clock-watching, which is precisely what I am doing, he starts to talk, in a surprisingly pompous voice which I hadn't heard before, and instantly never want to hear again.

"We have been looking for you, Karin. This morning, you were unusually late."

He even manages a sigh. I have never found Barry amusing but right now he almost makes me laugh. He is seriously expecting me to justify my lateness to him. *Wow! A Pussy with no authority trying to lord it... we are in that strange space where tragedy and comedy blur...*

I raise my eyebrows. "True. I was. You missed me?"

He looks confused, somehow surprised that I am not intimidated by his non-existent authority. "Eh... no. Or maybe, yes. No! Anyway, this project student of yours... what's his name..." *How did he cock up this time?*

"His name is Nick. What did he do?"

"Well, he came in this morning, but I sent him home. Actually Logan did. We did... He was drunk." *Barry, you couldn't get a toddler out of a sandbox... Of course it was Logan who kicked him out!*

"Drunk??!!" I feel my eyes widening at the hang-dog misery which is Barry's default expression.

"Yes, apparently he has a side job in a lounge bar...," he pronounced the words "lounge bar" as if he had never heard of the concept before. "They had a cocktail tasting for the personnel this morning. He was smelling of alcohol and was clearly unsteady on his feet."

"Right... Good job sending him home. What was he thinking? Drunk in a chemistry building? I will talk to Mark about it."

"Okay," says Barry, letting his shoulders drop again.

I walk to Mark's office with the timer in hand; a great device for terminating unwanted conversations. I just estimate how much time I believe I need with Mark, which is never more than five minutes, and set the timer accordingly as if the alarm means I have to do something important – which is sort of true; I *need* to leave. I am not the only one playing this trump card. Suspiciously many of the Lab 262 inmates hectically press random buttons on timers when we hear his keychain approaching – even some undergrads do it.

I knock on his door and he barks me in. Mark is behind his desk, hectically writing something on the A1 sized freebie calendar he got from a chemicals supplier.

"Crazy, crazy, so busy... and that within the semester break..." he mutters. "What can I do?..." he asks, seeming sort of pleased to see me.

"About Nick... Logan and Barry sent him home this morning because he was drunk, and it's not the first time it's happened... Last Wednesday he came in two hours late, which is par for the course, but so stinking of alcohol that he would stand out in a crowd of Irish football fans. Unsurprisingly, he emptied his guts in the lab sink. He destroys enzymes by forgetting to put them back into the freezer. He is an absolute monkey in the lab, which is an additional problem, beyond being pie-eyed. So, I've decided I don't want to have him as a project student anymore."

"Students..." he says in a weary way, as if struggling to martial an interest in such trite creatures. "You don't want to supervise him anymore?" It is actually more of a statement than a question.

"No, I don't."

"Okay, then Logan will have to do it." *Yikes, Logan will be soooo pleased...*

"Eh... What about just telling Nick that he is not welcome in our lab? Like firing him."

"Not possible. He needs to finish the project for his degree. It's part of his curriculum."

"Well, he doesn't seem to care too much about his degree, and if it turns out he does, he can do his project somewhere else... start with a clean slate."

My timer started beeping while I was speaking. Mark looks at me, eyebrows raised as if I just proposed we should switch to researching alchemy.

Mark shakes his head and smiles uncomfortably. "Logan will supervise him." *Poor Logan. He will hate me. But then again, he could also just say he doesn't want Nick as a student and then maybe Babette will get him... she will probably chain him upside down in the lab if he makes a mistake... despite all his ineptitude and sloth, Babette would be a cruel fate for hapless Nick...*

I go back to the lab with a smile on my face, thinking of Nick under Babette's dark wing. I walk straight to the small 37 degrees room where nine large Erlenmeyers with broth and *E. coli* have been shaking for the last hour. It's a tiny room without windows. It smells worse than a sewer but I don't mind any more, I'm accustomed to it. I take out two flasks and check from the outside if my bugs survived. Thankfully they did. I induce them so they – hopefully – start producing my protein, and close the door behind me.

Lucy sits at the bench with a piece of paper, pen and calculator. It's not an uncommon scene; Lucy can very often be found calculating enzyme concentrations for her experiments – she has a very predictable routine.

"Wow, you seem happy. I guess you had a good night?" she asks.

"Yeah, it was... hmmm... very... nice..."

"Oh God, you did not sleep with the Neanderthal, did you?"

"Unfortunately not."

"Tell me more..."

"Not here."

It is only four days since my head turned red and my heart beat faster as I read the Facebook messenger invitation. I had blurted out to Lucy, ignoring the fact that Logan was in the office as well, "He invited me out!"

"Who?" Lucy had replied with raised eyebrows and a slightly worried look.

"One of the three hot guys in this department!"

"Your definition of 'hot' differs from mine. Sex with semi-evolved primates doesn't appeal to me."

"You don't find the Scandinavian guy hot? He has a torso and legs to melt into. He has beautiful wild hair, chest hair sticking out of his shirt…"

I closed my eyes and inhaled deep breaths through my nose as if by doing so I could get a whiff of his bodily scent and testosterone. The only smell reaching my nose was from the methanol and chloroform mix. It's quite toxic for fantasy, and human lungs, so I came back to boring reality.

"No."

"It's not him anyway. It's the theoretical physicist in the office down the corridor."

"Thomas?!"

Lucy closed her eyes, rested her head in her right hand and shook it sorrowfully.

"Yes," I say, feeling much less thrilled. "What's wrong with that?"

"I just declined his invitation half an hour ago."

"He invited you as well?"

I tried to sound surprised, but of course he would have tried to date Lucy first.

"Yeah."

"At least he feels like going out then," I say gamely clutching for a bright side.

"Yes, but it is a bit cheap to try both of us."

"Maybe he doesn't know we work in the same office?"

"I'm pretty sure he does."

"Why did you say no?"

"He is closer to a human being than the monkey in woollen socks, but he still skipped too many millennia of evolution."

"He's one of the top three!"

"I suppose he represents one of the last three stages before *Homo sapiens* – that's as kind as I can be."

Lucy could read the excitement on my face. I was second choice – at best – maybe third of fourth, God knows who else in the department he asked. Too bad. The prospect of going out with other men was what I needed.

Lucy looked at me as if she totally understood.

"Where are you going? Ceilidh dancing I presume? That's what he proposed to me at least."

Wow, he really knows how to make a girl feel special…

"Yes."

"Ceilidh is not dancing. It's an invention of the Scots; a sport consisting of people whirling frantically and tossing each other about while folk music happens to play."

"Well, at least physical strength counts for something… and there is someone instructing everyone to keep the whirling and tossing vaguely synchronised…"

"Yes, yes, it's a good place to go 'dancing' with partners who can't dance which, I presume, a theoretical physicist is."

"Yep."

"What will you tell Daniel?"

"That I'm going to the bar with you."

"Make it a ceilidh, otherwise he might get suspicious. A normal bar night does not make you leak three litres of sweat."

Logan had been packing the autoclave and had overheard the whole conversation. He walked to the office door, turned around and said: "Can I give you a disapproving look, Ka?"

He was right. What I did was wrong. But it was good to be held on the dance floor by two unknown strong arms. It was good to have the feeling of being special and wanted; the warm feeling I get in my stomach if I think about his dark brown eyes, filled with emotions, and his messy black hair… His eloquent and smart words with an exotic accent made me melt like a little Belieber being invited backstage. But he is much too intelligent for me. Doesn't matter…

Lucy takes our communal pack of cigarettes out of the drawer and we head out for a smoke.

"Did he gnaw on your ear?" she asks, imitating a rat.

"No, he didn't gnaw on any of my body parts."

"You kissed?"

"No, only nuzzled between each other's legs," I say, sarcastically and sniffing like a dog.

I am telling Lucy about the evening before, how Thomas had talked to me, looked at me and led me over the dance floor. The strange character of the Scottish dance had actually enhanced the romance. I had felt good. It was a rare feeling, an alarm call; how joyless my life has been for so long…

"Did you tell him about Daniel?"

"I did, but he didn't seem to mind too much. And Daniel wasn't suspicious either about my whereabouts. So, all good."

When I came home and got into bed Daniel had said: "Why are you so late? I've been waiting for you."

He leaned over and wrapped his arms around me in such a way that I couldn't move. I didn't give him an answer and waited until his breathing was heavy. I moved his arm off me and rolled my body to the other side of the bed. I don't know if it was guilt or distaste for the touch of Daniel, but either way I should have felt bad. Yet I felt great.

The late evening sun is projecting shadows of fluttering leaves on the wall of the lab. It gives the feeling of summer, which it is. It is Friday evening and there is only Lucy and me in the lab. I'm preparing some buffers while telling her about what happened with Nick; that Logan got to supervise him instead and how thrilled he wasn't by this. Thankfully he didn't blame me for the swap, just Mark.

Suddenly the sun is not only projecting the shadow of leaves on the wall but also the shadow of a man. My heart stops and my body stiffens. It is easy to scare me. As a kid I didn't want to sleep alone in a bed. I was even scared to go to the bathroom in the evening as I had to pass the glass front door. With the light on in the hallway and the pitch dark outside I knew anyone passing could glimpse me while I could never see them.

"I didn't want to scare you."

It is a low voice. I'm still staring at the wall. From the voice and the posture I know it's him. I turn around and he smiles at me with his open, intimate gaze. He's even more handsome now than he had been at the ceilidh, with his shining eyes and strong jawline – so manly.

"We didn't hear you enter."

My heart is beating fast and my voice is trembling.

"Sorry, next time I will make more noise."

He really does look as if he regrets sneaking up on us.

"You are still working?" *No, I am just pouring solution A and B together for the sake of it; I can't get enough…*

"Almost finished."

"I wondered if you fancied going for a walk, to Arthur's Seat in the evening sun." *I would love to!*

"We won't make it in time before sunset." *Stupid girl! We would easily make that. It will be light for another three hours or more. Now he thinks I'm retarded or don't like him.*

He looks around the lab, thinking carefully about his next question.

"Maybe your boyfriend would like to come along? I would love to meet him?"

"No!" I say, way too loud. *What a dreadful idea!… That was witty.*

Thomas bites his lower lip and smiles at me in the super-sexy way he did repeatedly yesterday evening.

"No," I repeat, much more casually. "You two are probably too different. Plus, I planned to go for drinks with Lucy."

He looks disappointed.

"Another time?" *I shouldn't, I really shouldn't, I musn't, I won't...*

Lucy is looking at me, more curious about my answer than he is.

"Sure. What about tomorrow?"

"I'll give you a ring in the afternoon, okay?"

"Okay."

He walks out of the lab.

As soon as we hear the door close she blurts out: "You like the Neanderthal!"

I struggle not to be obvious. "He's okay."

"What about Daniel? That dude you live with?"

"I guess I will break up with him."

"About time."

Lucy's expression is wry, very knowing. She certainly isn't overloaded with moral fibre when it comes to infidelity.

We head to the city, deciding to have a beer in the Royal Oak, the Scottish pub close to South Bridge. We normally don't go here but tonight I feel like listening to live Scottish folk music, filled with emotions. *Am I really going to break up my (very) long-term relationship – just like that? When did I become a fickle hussy?*

Part III: Year 2

Chapter 19

I close the front door of the flat and start to run. I charge through the residential area east of Gorgie Road and then through an industrial park until I hit the cycle lane to Leith. I follow the tree-lined path which lies a few metres above the city and then I take the steep muddy stairs down to the wide stream known as Water of Leith. I run the stairs up and down till the muscles in my thighs burn. I run as fast as I can through the forest surrounding the Water, over beautiful little wooden bridges. It is the land around this stream which I will always remember as the most enchanting part of Edinburgh; the hanging trees, the verdant little hills rising steeply from both sides of the water, and the unexpected calm – seemingly far away from the hustle and bustle of the City in which it lies.

I am often to be found here, especially at weekends. Today it is exceptionally quiet, even for the early morning which is normally the preserve of dog owners. There is a layer of fog hanging over the river. I follow the stream and run as fast as I can till the trees and greenery come to an end and the river makes its way through Dean Village – an exclusive part of Edinburgh resonating with picturesque history. I turn round and run the same way home. It is only about five miles there and back but I am covered in sweat and completely out of breath.

Daniel is standing in the kitchen, wearing boxer shorts and a T-shirt. He pours boiling water on top of the layer of dry coffee powder in the French press. He takes a bag of bake-off breads from the shelf and turns on the oven.

"We need to talk," I say, standing still in the doorway, breathing loudly.

"Okay," he says, placing the coffee and two mugs on the table.

© Springer International Publishing AG 2017
K. Bodewits, *You Must Be Very Intelligent*,
DOI 10.1007/978-3-319-59321-0_19

We both sit down on the benches that I had painted red the weekend after the plasmid fiasco. Like the running, the painting was some sort of meditative exercise to clear my mind. This noble quest for doctorate glory was supposed to prove I have a brilliant mind, yet it has rendered me unable to think freely and confidently. But at least I know it. After my run, life seems a touch clearer: "Daniel, I think it's better if you go back to the Netherlands."

"Why?"

"Because I am not sure I still want to be with you."

He looks at me genuinely surprised, I think.

"What?" he sort of whispers.

We both fall into silence, like two strangers with nothing to say to each other. As we sit here, I imagine discussing everything I feel or have not felt during the six years since we met, but I don't: How can I tell him after travelling together – illegally hitchhiking into Tibet, celebrating New Year with a village family in remote central China, sleeping with nomads in Kazakhstan, all the times we camped in mysterious wilderness – that it was the experience that made me happy, not him? How do I say I never intended to stay with him for long, that I was not in love with him, that it just dragged on without clear reason? Am I supposed to say that I hate the *way* he never seems to succeed at anything, more than the *fact* itself? Or that I do not believe he can ever hope to find happiness by doing nothing? How can I express the truth without hurting him horribly, perhaps for life?

After a while I drop the bombshell, "I've booked you a flight, you leave tonight."

"Tonight? So... you already made your decision?" *Eh... yes. Please tell me we're done here?... But I know, nobody ever gets to just skip away from a long relationship, even if it is long dead...*

His tone of voice suggests that he doesn't like the fact that I made the decision alone. It's casual enough, like the way you might react if your partner did not involve you in picking the next movie to watch.

"Yes." That is all I can think to say by way of elaboration.

He looks flabbergasted, as if he cannot quite comprehend what is going on and didn't expect it despite the millions of signals I am sure I sent. From the expression on my face he can probably read that I am deadly serious. I know he needs time to come round. I know I want this over.

"But I was planning to marry you. I was even making plans to ask you!" *I was afraid of that. And I wouldn't be good for you...*

"I'm sure you will find someone else."

"I don't want to look for someone else." *Then don't. Let me go! PLEASE...*

Not long ago he told me that he was very happy, in contrast to many of his friends. He was in the luxurious position of not having to hunt for girls – a process he finds very draining. When he spoke those words, I felt a twinge – I already knew he would be rejoining his friends in that draining activity. But I didn't know it would be so soon.

"We have done so many things together. How can you throw all of that away?" *The thought of doing even more together certainly helps. Oh God, this is so* over.

Tears are welling up in his eyes.

"When we're travelling we get along well. We can spend hours together in trains, buses or in pubs. But don't you realise that outside of those times we don't flow well together? There is no passion."

"What bullshit! There is passion. After all these years we still have sex – often. And we do things together."

"We have sex every other day Daniel, not more, not less. It is like a duty, nothing else."

He doesn't like the topic, he never did. How often I have pointed out that it was not good enough for me, that I wanted something different. I veer away from that eternally touchy subject: "Besides, passion is not all about sex. It is something different for me as well. I want to be longing to get back to a partner I respect. But if I have the choice I seem to opt to do stuff without you. I stay late in the lab, and try to go out with others."

"Are you seeing someone else?"

"No," I lie.

"So, there is still a chance." *No.*

Daniel begins to cry proper. I go to his side of the table, give him a kiss on his electrostatic hair and wrap an arm around him. I am not sure if I ever saw him cry before, maybe when his grandmother died but even then I'm not sure.

"It will all be fine," I whisper, while caressing his back.

He tries to kiss me on my mouth and I decline at first. When he tries again to move his mouth close to mine, I decide to reply to his kiss. *After so many years, one more kiss doesn't matter.* With his mouth pressed on mine I feel his anxiety. He is afraid to be alone. So am I.

Slowly, I push him away from me, signalling that kiss was final. I walk to the oven, get the bread out and place it on the table, but neither of us eats. We just sit. Daniel is still sobbing. I feel defeated.

"I will help you pack," I say, and stand up from the kitchen table.

I am sitting on the window sill, watching the city slowly come alive. The rooster in the City Farm on the other side of the road is crowing when Daniel comes in and asks me for a cigarette. Smoking so early is usually not his thing. We both light one and stare at the people passing on Gorgie Road. We gaze at the old buildings, admiring their beautiful architecture, but we don't speak. Thank God there is music in the background to break the otherwise painful silence. I wonder if Daniel will miss the view, which he has arguably been savouring far too much in recent months. Perhaps he will put it all behind himself quickly, forget Edinburgh the moment he leaves it. I hand over the whole pack of cigarettes; I guess he needs it more than me. I pull a large suitcase out of the wardrobe in the living room.

"What about my boat?" he asks.

Oh yes, the damn boat. Waiting to be sailed; sleeping, somewhere at a sailing club close to the Forth Road Bridge. Daniel has a small, one person sailing boat that, like its owner, exists in permanent passivity... lying around somewhere. Daniel insisted on his parents dragging it along to Edinburgh for Christmas. He envisioned golden opportunities to pick up his old hobby in his new home. It never happened.

"You need to pick it up by car later," I say and walk to the door.

"I guess so."

"I will be in the lab for a few hours," I say, though I have no idea where I might go.

I do go to the lab; let the time dissolve in experiments. I return an hour before he is likely to leave. I need the confirmation that he is actually gone. When I enter the flat I see the large suitcase standing ready at the front door, which is a comforting sight. Without clothes on the floor, the flat looks empty as if Daniel is no longer a presence here. And it is just the ghost of Daniel sitting and smoking at the window sill – precisely where I left him this morning. He doesn't greet me. He seems to be thinking, or dreaming, I can't tell. He looks desolate for sure.

"You know what scares me most, Ka?" Daniel says, still looking out of the window.

"No," I say.

"That you won't pull back!... Once you have made the decision to separate from me you will stick to it, even though you might miss me and might want me back."

He starts to cry again and his shoulders are rocking. This time I decide to stay at a physical distance. Between sobs, he says: "By taking me back, after a week or even a month, you would have to admit you made the wrong decision. That you failed to dump me."

He looks me in the eyes and adds: "And Karin doesn't fail. She never fails."

He walks out of the living room and opens the front door. When I hear it closing I know he is gone. I start to cry.

Chapter 20

I can't concentrate, and I can't work. It's nobody's fault but mine. I dumped Daniel over a week ago and I feel desperately sad.

I am practical, I know it will pass. But that knowledge is of limited solace. Of course lots of people get into this situation, especially my PhD peers, failing to keep relationships alive over time and distance, and during penury while ardently pursuing another passion. Knowing that doesn't help either. I feel small and vulnerable and lost abroad at the end of a relationship. Life still hurts and looks evermore like a pointless chore we tend to because we have nothing else to do.

I wish I could lose myself in my PhD, yet non-stop all day long I try to figure out ways to avoid the lab as if I would thus avoid this overwhelming bleakness.

"You look amazing! This whole being single thing suits you well," Lucy says, with obvious sarcasm, when I enter the lab just before midday.

Logan starts laughing on the other side of the bench as if this were the best joke ever.

"Oh, don't laugh," I say, with a weak smile.

"Sorry, sorry. I shouldn't have, but you don't look fit."

"Ibu plus coffee breakfast doesn't make anyone light up like a Christmas tree."

"Go home," Lucy says. "Relax for a few days, then come back."

"I do not want to relax and I do not want to be alone in my flat."

I walk over to KB House and slip into my running shoes. I run down the campus hill and turn north at Dalkeith Road to the steep incline leading towards Holyrood Park. My body, hitherto unaccustomed to chain-smoking

© Springer International Publishing AG 2017
K. Bodewits, *You Must Be Very Intelligent*,
DOI 10.1007/978-3-319-59321-0_20

and heavy drinking, has become intimately acquainted with both in the last few days, so now it pays the bill. My lungs hurt and my liver aches, but as time passes, I imagine the organs cleansing and replenishing. I run through the park along the foot of Arthur's Seat passing a few ponds sitting pretty in the midday sun. Lots of office workers have brought their lunches out to enjoy the rare weather; probably not as many as you would expect based on the size of the town, but this is Scotland, people are simply not hung up on weather.

I am wearing a headset, listening to music as loud as my ears can endure. By the time I reach the Palace of Holyroodhouse on the other side of the park, I feel my head is emptying – clearing – for the first time in days. I don't feel anything apart from tiredness in my legs. I run back along South Clerk Street, much slower than before, and walk the last bit up to campus. After my shower, I see a few of my colleagues having lunch in the little garden behind the canteen. I'm still not hungry. And I notice I am still not ready to socialise because I decide against joining them.

Back in the lab it is just Barry and Logan working on the new chemicals database. Barry did not get a summer student to supervise but he got responsibility for the database instead. He is so pleased with this assignment that his arms seem to hang even lower than before. Logan hands me a sheet of barcodes.

"Why don't you help out?" he says with a supportive smile.

"A perfect task for today," I reply, taking the sheet out of his hands.

For hours thereafter I lift bottle after bottle of chemicals off the shelf above me, register what is in the bottle and how much is left, and then I stick a barcode on it. While doing this, it sometimes seems that the number of bottles still to be done is increasing and, at best, never shrinks. I estimate that in the unlikely scenario of combining forces, it would take us at least another full day to finish. In our lab, we have so many chemicals which no one ever uses, and many of them are toxic. They have been sitting here for years upon years; many presumably purchased in the days when the lab was still run by Prof. Gilton. Some stick to the shelves, many of the bottles have faded labels and decayed lids. This makes it difficult to tell what is, or has been but, worse, it also makes you wonder how much of this potentially harmful – even carcinogenic – stuff is seeping into the air around us. I collect all the unidentifiable chemicals and the ones which have long passed their expiry date and place them in a plastic box for disposal.

Late in the afternoon a bottle chances through my hands with the name of a ^{13}C isotope-labelled amino acid which rings a bell. If I recall correctly someone from another research group wrote a round robin email, just this morning, asking to borrow this particular compound. I take the bottle to the office and compare the name on the label with the text in the email. It is identical. Felix, working in Lab 018, wants this. Hurrah, yet another excuse to get out of the lab. I don't risk phoning in case Felix is not around or already has the chemical.

I pass the chemical stores and enquire where I can find Lab 018? "Just around the corner, first door on the right," says the young guy pointing in the direction of the main corridor.

It's a clear direction, but when I open the door, with the right number written on a sign next to it, I am in an office. *And what an office! Forget about fair distribution of goods! The roof isn't poured in gold, but that's about it...* From just a step inside the door I gaze in wonder for a few seconds. *Our lab would fit in here at least four times over – and this is just their office!* There is a corner with sofas, a large conference table which could seat twenty people, a projector in the middle and a several desks with computers and printers. The walls are decorated with scientific posters and framed covers of *Nature* and *Science* magazines. *If you make it onto those covers, you rock!*

I've clearly entered a hallowed working space, inhabited by one of the gilded robes of the university; a Professor who has unequivocally made it. Just next to the door there is a kitchen block with a large coffee machine on top. Mark doesn't want us to have a coffee machine in our office; curiously, not even if we purchase it ourselves. I guess he doesn't want to forego the soliloquising opportunities of coffee breaks at KB House. A machine would save us all money; in my case about 5% of my humble stipend. The thought that I pay to listen to Mark's monologues somewhat rankles.

There are a few people chatting at the kitchen block so I venture to ask who Felix is? In so doing I take a step forward whereupon a bald, overweight, middle-aged man, holding a pack of crisps and looking like Homer Simpson in a too short T-shirt, notices my presence and tells me: "No chemicals in the office, young lady."

"But I'm looking for a guy called Felix," I say.

Homer Simpson looks around the circle of people he is talking to, all much younger than him, expecting someone to enlighten me.

"Probably in the lab," says a tall, blond guy, gesturing to a glass door.

"Thanks."

I take a few steps in the direction of the door. "Wait. You can't walk with chemicals through here, just walk around," says Homer Simpson, though with a friendly tone, indicating that I have to enter the lab via the main corridor. *Pedantic Flabby.*

I obey silly Homer and walk out through the door I came in.

The lab turns out to be even larger than the office and, presumably due to its glass roof, boiling hot. There is heavy metal music playing, medium loud, and everyone is working in fume hoods, wearing a lab coat and safety goggles. Except during the practical classes I teach, I never wear safety glasses. It is not necessary in our lab although we undoubtedly break the health and safety regulations by working with carcinogenic solvents like chloroform unprotected. However, we have nothing that can actually explode, or at least that is what I assumed till a few hours ago; but with all those unknown chemicals freely evaporating or forming dust I am not sure anymore. Homer's lab is so different; people are doing hard-core chemistry and for sure they need protection. It's industrious and scientific and really quite hypnotic and uplifting.

A guy washing his hands at the sink next to the door greets me and asks if he can help. I explain I'm looking for a guy called Felix and he nods in the direction of a fume hood at the back. "The red-haired guy," he adds.

Without saying anything, he points to a box with safety goggles for visitors on a table next to me. I pick the only red plastic goggles, place them on my nose and cross the lab. I pause next to Felix.

"Just a sec," says Felix while exchanging one tube for another and catching a transparent liquid. There is a large colourful column hanging on the stand above the tubes; chromatography, I think.

"Yes," he says looking at me only briefly so as not to lose visual contact with the dripping liquid.

I put down the bottle with the amino acid and proudly announce, "We had it."

"Ah, cool," he says, quickly reading the label and exchanging the filled tube with an empty one. "Do you have a minute so I can take out what I need?"

"No worries. I have time, lots of it."

I sit on a stool and watch for a few minutes how skilfully he keeps on exchanging tubes, as if he has been doing it for years. It is quite dreamy in this lovely lab watching the mysterious liquid drip into tubes to a soundtrack from Opeth, Swedish heavy metalists.

"You like the music?" he asks after a while.

He must have seen me dreaming away at my mirror image in the fume hood window. "Not really, but it fits the scene nicely."

"What do you like?"

"Rock, jazz, blues… depends…"

He exchanges the tubes one more time and closes the valve above the column. "Finished."

He turns towards me and I see shiny blue eyes behind the safety specs, and a friendly, freckle-covered face. "You don't have a fraction collector to automatise this process?" I ask, pointing at the 100 or so tubes of liquid he manually collected.

He places a hand on my shoulder and grabs my under-arm as if we have known each other for years. He bends his knees, loosens his hand and makes a waving gesture towards the samples as if he is presenting them: "Why would we need that? There is ME…"

He looks playful, entertained. He grabs the bottle with dried amino acid and I follow him to the scale on the other side of the lab. "I am a postdoc, much cheaper than a fraction collector," he says, smiling. "In fact for my boss it's free; I'm funded by a European scholarship to work here."

He takes a small piece of paper folded in the shape of a tray and carefully places it on the middle of the scale. With a spatula he takes a spoonful of amino acid and weighs it. "You mind if I take a bit more?" he asks.

"Go ahead."

"You are an angel."

"You are the first person to tell me that."

I do not feel like an angel. I feel cruel and broken. He smiles again. He is a happy bunny, a person that lightens everyone's mood.

"Can I give you something in return?" he asks, while looking around the lab and, I guess, realising this is a rather stupid question.

I could have suggested "a real and meaningful PhD experience in this nice lab," but I don't want to sound as desperate as I feel. "Like pipette tips?" I joke.

"Yeah, you want?"

"No, I don't, but I wouldn't mind a coffee from that amazing machine you've got in the office."

He looks slightly surprised. He has possibly never worked for a threadbare underpants-type of guy who does not have a coffee machine in the office next to his *Nature* cover pages. He possibly doesn't know that the PhD world contains deserts next to verdant grasslands like his blessed lab.

"Sure, come," says he, steering me by the elbow towards the glass door.

"Can we go through here?"

"Of course, that's what doors are made for."

"I got a bit scared of Homer Simpson just before, telling me off; he said not to walk in the office with chemicals."

"Homer Simpson?!" he repeats, obviously amused. "That's the boss, Professor Walker… He's not scary, just fat."

I look at him in disbelief. Is he really telling me that the Homer Simpson guy with a white, too short T-shirt, jeans displaying a builder's bum and thick clumsy fingers shining with the sunflower oil of his crisps, is the same person featured on the cover of *Nature* and *Science*? Is Homer the person who has at least twenty-five PhD students and postdocs working for him?

"You are joking," I say, pausing in front of the glass door.

"Nope," says Felix, steering me by the arm again to make me move.

We walk into the coffee corner but this time there is no sign of Homer. There are just a few people working at the desks facing the walls. They all look like they are engaged in serious work. If my screen were visible to everyone in the room, including Prof. Simpson, I would also refrain from opening social media on my desktop…

"Shit, no water tank," says Felix, placing the coffee mug on the roaster and checking the sink.

"William's got it again," says a guy at one of the desks closest to the block.

Felix smiles, sighs and shakes his head playfully. "Come."

We walk through the glass door back to the lab. Felix opens the door of an office cubicle separated by large windows from the lab space. It contains a very large U-shaped desk going around the office. It is separated into four generous work spaces. There are two caddies with golf clubs, a golf ball target under the far end of the desk and a few balls spread over the floor. *Someone must be playing inside, wow!* Sitting in front of one of the computers is none other than red-haired Pippi Longstocking-Heidi crossbreed, whistling the same tune she always whistles in the corridor. There are two other people in the office, both wearing headphones, possibly to block out her whistling.

Felix rests his hand on the shoulder of a sporty, black-haired guy drawing complex chemical structures on his computer. He waits until the guy takes off his headphones. "William! You happen to have the water tank for the coffee machine?"

His desk is tidy, everything perfectly aligned and placed just so. Age-wise, William is probably too old to be doing a PhD, a postdoc I suspect.

"Yup," says he, turning around in his office chair to face us. His demeanour is either cheerful or overly strict. For me as an outsider, it's hard to tell, but quite amusing to watch.

"This young lady dropped by to give me the amino acid we need to build the next station; I would like to give her a coffee in return."

William takes his keychain out of his jeans and uses it to open his desk drawers. I can see scientific papers in there, divided by splitters in alphabetical order. At the back of the drawer there is a large water tank, pushing all the papers to the front. He takes it out and hands it over to Felix.

"Please return it when you're finished," says William.

"Sure," says Felix and slaps him on the shoulder in a friendly way.

"Can I ask why you keep the water tank of a coffee machine in your drawer?" I ask.

"Because the last person using it didn't clean it; only this way do those guys learn," says William in a very irked tone, as if dealing with a small child, and nodding in the direction of the lab. *Control freak.*

Felix shuffles me out of the office. "William is loopy," he says when we are out of earshot. "His research team is numbering the meetings World War I, World War II, World War III…"

While filling up the tank, he adds, "He's a good chemist though, great guy."

He hands me coffee smelling of freshly ground beans. "If you want another one, just help yourself," he says, clearly not intending to return the water tank to William.

"Maybe tomorrow," I say, sort of hoping that I have finally found a free coffee supply but also wondering if maybe, just once in a while, academia still works and is a fine thing – bizarre boss, neurotic employee, sly banter, mad music and general eccentricity in a state-of-the-art happy lab resounding with intense industriousness. This is what I thought I was coming to Edinburgh for…

"Any time, just join us," says Felix.

I wish I could. I really wish I could. "I will come back to return the mug," I say.

"Keep it for next time, it's mine."

Out in the main corridor I hear the same yawning noise I regularly hear in the early evenings; one of the senior lectures practicing bagpipes in the large lecture theatre. Given my general bleakness about life passing by unremarkably, right now that bagpipe racket sounds like a marching dirge as I trudge from what-should-have-been to what-sadly-is.

Chapter 21

"You don't eat?" Thomas asks with a worried look that make his brown eyes look even sexier than when he smiles.

"I'm not hungry."

There is a full plate of beautifully decorated Spanish tapas in front of me that I would normally love to eat, but today it makes me feel sick. I haven't eaten anything substantial since Daniel left. I get my calories from alcohol.

We had finally parted, for good. It is a feeling of relief, like the one you might feel post cecum removal; the appendicitis has been alleviated but, at the same time, it hurts. What hurts me most is that I hurt him. The desperate messages he sent me during the last two weeks trying to win me back, begging to talk it all out and so on. The pain in his words and his trembling voice. All his emails and voicemails end with similar content; that he is not sure how it all happened but that I am definitely not the same person as I had been before the PhD. Three days ago, he said I should go and see a general practitioner to get checked, something must be wrong. Two days ago, he suggested I go and see an endocrinologist – clearly an imbalance of spooky female hormones has driven me to an irrational decision. Yesterday he proposed a psychologist. I'm half-curious what specialist he will suggest tonight. Lucy and I speculated a cardiologist, which wouldn't make any logical sense, but it would fit the dramaturgy. Next week it might well be a brain tumour causing my foulness, though Daniel could just as well recommend a tarot reading. I feel sorry for him but I also know that his desperation will yield to anger, misunderstanding and years of throwing mud in my face. Such is life when you dump someone who does

© Springer International Publishing AG 2017
K. Bodewits, *You Must Be Very Intelligent*,
DOI 10.1007/978-3-319-59321-0_21

not want to be dumped. Sometimes I yearn for that mud in my face ASAP; it will be one step closer to the end of the end.

I am too sober. Our story hangs over me like a cloud of gloom. Sexy, it is not. I can't talk to Thomas about it anyway; I would bore myself as much as him. He knows it is over now, and despite the fact he never asked, he probably has an inkling that the dumping of Daniel so soon after the ceilidh is not entirely coincidental. Thomas gave me license and confidence to pull the trigger. He symbolised the fact that I could score better, or at least differently, or at least someone whose sloth doesn't make me want to howl at the moon. True, Thomas is someone who makes my stomach heat up in a manner it has not done since my teen years, but he needn't fear it; I'm not looking to pick out curtains with him. I just want to play while time passes.

Today my stomach does not boil. I feel empty, exhausted and wonder if it would not have been better to go drinking yet again with Lucy. Thomas had phoned the very evening Daniel left. I had told him as serenely as I could – between stifled tears, and therefore probably not very serenely at all – that I just needed a bit of time, I'd get in touch soon. He has been a patient gentleman and so, after catching sight of him strolling on campus this afternoon, I dropped by his office and finally accepted his dinner invitation. And here we are, mostly just staring at each other and not knowing what to say.

"Shall we have a coffee at my flat?" Thomas asks.

Some mindless sex with Mr. Hotty? I'd be crazy to decline. I want to get out of this place with the Bueno Vista Social Club blasting through the stereo and only leaving me cold... But, er, coffee!?...

"You have wine as well?"

"Sure," he says, smiling and showcasing perfectly aligned white teeth.

He waves to the barman and pays the bill. He is just trying to be gentlemanly but I am transported. With Daniel I always paid the bills and it got my goat eventually.

We still don't talk much on the way to his apartment. It's faintly awkward but it's better than me prattling on like a misery guts about my big break-up. Lucy had warned me in the office today not to ask anything about the mysterious theoretical models he is working on. "Give them a finger, and they'll take the whole hand: if you show interest in their matrices then next thing they are enthusiastically drawing formulas on the windows. And then you sit there, utterly bored and wondering if the Greek sigma symbol turns him on more than you do."

She had rolled her eyes in disbelief when I said that a Russell Crow-clone covering glass with complex algebra would turn me on, though I feared most physicists would use unromantic paper instead.

"Totally agree with Ka… so hot," Hanna had chipped in, raising her head from her lab book.

"You both have strange sexual fetishes," Lucy said.

"What turns you on then? 'Lucy…I have a première édition de 'Batman the movie' Tu veux le voir… together?'" I said with a low voice.

All three of us laughed. Logan placed his hands over his ears. I added, "Give me a guy whispering complex formulas in my ear… that makes me much happier!"

"It makes you feel stupid," said Lucy.

Hanna didn't care, "You don't have to follow it, do you? Just watch him passionately, but hopelessly, explaining something complex. And then you have awesome sex afterwards."

Logan left the office as soon as he heard the word "sex." He is unnerved by three girls talking intimately and crudely. But maybe female scientists are weird? Maybe male scientists are weird? Most people do not want to be scientists, yet scientists cannot imagine wanting to be anything else. We possibly nurture each other's weirdness. I don't know any more; scientists are my world.

We often talk about our sex lives as a fun game, sounding like wised-up students while we toil through our PhDs. Mostly it works, but later in the day, such as right now, it can feel remote because in truth we are older and life has gotten heavier – unbearable at times.

Despite having sexual fantasies set in Oxford-like movie sets, tonight I would prefer a drink over algebra. I find myself hoping that Thomas doesn't turn out to be an ultra-nerd. Suspicion is ignited when, at the last set of traffic lights by the Royal Botanic Garden, he watches me for a long time. It is definitely weird.

"You feel like coming home with me?" he finally asks.

"I guess I do… As long as you are not planning on watching *Batman* with me."

Thomas laughs in a charming way that relaxes me.

"Why did you come up with that?"

"Just a thought going through my head."

"Strange girl you are. Do you like it in Edinburgh?"

"Yes, I do. I like the city, but I'm not sure if I like my PhD."

"Mark gives you a hard time, doesn't he?"

"He does."

I am shocked to feel tears welling up. I don't know if it is because of the question he asked or because of the emotions of the last few days. I need a cigarette or a bottle of wine to suppress my feelings, but I don't have either on me.

During the last few hundred metres to Thomas' flat, located in beautiful Stockbridge on the north side of town, neither of us speaks.

"Here it is," he finally mumbles.

We climb the stairs to the top of a three-storey building, which is cold and reeks of detergent. Thomas sticks his key in the front door then turns around and softly touches my cheek with his hand. *Tender or weird? Or both? And do I care? Hm... This is going well...*

I step into the flat behind him and see that I am already in the living room. The ceiling is high and decorated with angels, and the evening view from the windows, over the Firth of Forth is stunning; so different from Gorgie Road. Soft blues music sets the tone while two guys play chess at a large wooden table. Both lift their heads to greet us and I instantly recognise their faces; PhD students at the Chemistry Department, working in different sections from me. I see them often enough playing cards or pool in KB House on Friday nights. *Shit.* You might just as well announce on the flat screen at the entrance hall of the School of Chemistry, "Person A has agreed to have intercourse with Person B." Men might feel proud in such a situation and want the world to know they got laid. But women like to feel it's special and nobody else's damn business for now – bang goes that idea. *Why did Thomas not mention he lives with these guys? And why on Earth did I not suggest we go to my flat instead? I'm really not on my game, so out of practice, is there a course on casual intercourse etiquette?...*

The guys look far too curious and stare at me for far too long. I feel my face getting red and am actively wishing I was not here.

"Hey Karin, how you doing?" one says.

He has red, curly hair and glasses which are much too large for his face. The other one has long black hair tied to a ponytail and a belly suggesting he appreciates McDonald's cuisine. I have never talked to them, but they seem to know my name. *Maybe Thomas mentioned me to his flatmates before? He did not mention them to me. Is that slick operating or lowdown games?...*

"Hi, I'm fine," I say, probably sounding as insecure as I feel and look.

Thomas senses my discomfort and steers me towards the kitchen.

"White or red?" he asks

"Red."

"This is a good one," says he, taking a bottle with a bull on the label from the shelf above the fridge.

I wouldn't really care if it's methylated spirit. I just want to get away from the here and now. He takes my hand softly and gently pulls me to the other side of the corridor. We enter the small room in which he has obviously been living for a while and sit on the bed side. There is a bit of space between us but when he moves, the hairs on his arms touch me. For the first time today my stomach seems to warm up. It doesn't get super-hot as it did the evening he entered the lab but, hey-ho, it's nice to feel something other than abject misery.

I look around the room. A bookshelf filled with study books and a few novels, a desk, a wardrobe, a sink, a frame of a bike and tools and cables everywhere. *Why do guys always make their rooms look more like a garage than anything to live in?*

My eyes are drawn to a picture stuck on the mirror over the sink; Thomas with a girl. They hold hands while hill-walking, and they look very happy. I notice another picture on the desk, the girl alone. I survey the room carefully and note there are pictures of her everywhere. It's getting a little nerdy-weird, faintly creepy and even a tad Stephen King…

Wait, I know her! She works in the Chemistry Department and gave a presentation the other day. Lucy told me she is his ex-girlfriend. Wow, he is obsessed with her! Or is he back together with her? And, if so…do I care?…

I take a large mouthful from the wine glass which Thomas had pushed into my hand, evidently sensing my discomfort. *When a man plans to lure a woman back to his den, surely he knows better than to have it looking like a shrine to his ex? I mean, this hardly gets the juices flowing… Is Thomas weird?… Or plain thick?…*

He comments, "She is my ex."

"Mm, would never have guessed… Why do you have so many pictures of her?… You know, sort of everywhere?…"

Does Daniel have as many of me? God I hope not, he'll never find someone else if he has.

"Because I still love her," he replies.

Wow, again, how to make a girl feel special…

I sigh and all thoughts of romance vanish. I think the colour in my face disappears. A certain deadening of the internals certainly occurs.

Meanwhile, Thomas looks sentimental, as if he is about to cry. I stare into his big, brown eyes which are becoming watery. *Oh god, please don't cry… I can't handle that. I am not a social worker, I am a scientists dealing with my own mess right now – such as a grieving ex!…*

I observe him in silence. Soon I am struggling to suppress a smile. I feel a sense of relief. I feel safe, knowing that, like the evenings before this one, there is only drinking on the agenda for me.

"Why do you smile?" he asks.

"Because this is quite bizarre."

He actually looks surprised by that answer. I feel compelled to elaborate but I had thought it might have been more obvious than the colour of milk.

"I could imagine and accept a lot. Like… you turn out to be married or have a kid. That there is a picture frame on your nightstand which reminds you of your wonderful wedding day, or a holiday from days before you took other women home... A picture you would quickly shove in the drawer before you press your heaving body on top of me. I could imagine that we would have sex until the early hours, and then I need to hurry out of your flat before she comes home... But I could not – did not – imagine that your room would be a shrine to your ex? I just didn't see that one coming, Thomas..."

He looks away.

"I was very happy, standing on Arthur's Seat with you, watching the sunset, dancing at the ceilidh. But I can't get her out of my head."

He picks up one of the picture frames. I see desperation in his eyes and, at another time, I might have felt sympathetic. There is an honesty, perhaps even a purity, in his calm, broken manner. Alas, I am carrying too much emotional baggage already to take on his shit. I want to get out of his room, the shrine-cum-cell where he must spend hours upon hours brooding about her. I want away from the pictures of her and the happy couple, and away from this miserable moment. The only thing I want to take from this room is that bottle of wine.

"Let's go," I say, standing up.

He looks surprised by me stuffing the bottle in my handbag, but he doesn't say anything.

"You don't want to stay?" he asks. *Even chalk formulae on the window couldn't get that motor going anymore, dude...*

"No. There's a flat party at Mareike's place, I want to go there."

Thomas doesn't look as if he knows Mareike. "She works in the Johnson group. Brown curly hair, German."

It doesn't seem to ring a bell. "Are you invited?" he asks.

"Nope," I say, opening the door of his room.

I walked much faster than necessary in the direction of the front door. Thomas follows behind. His two flatmates are organising the chess pieces to start another game.

"We're going to a party, you guys want to join us?" I ask, hoping they will say yes and save me from a super-awkward half-hour walk with Thomas.

They look at me slightly surprised. *Yes, that wasn't enough time for Person A to have intercourse with Person B and as you possibly know it is because Person B can be pretty bloody weird...*

"Eh, no thanks, we're just starting another game," says the long-haired guy, the other nodding along. *Yeah, such fun guys in this house...* I vaguely hope that Thomas will take the chess option, but we end up leaving together.

We walk through the streets, still much quicker than necessary. I pause at a newsagent's to buy cigarettes. I light one straight away and hand the pack to Thomas. Just as we pass the train station to head towards the south side of the city my phone rings. As soon as I press on the green horn Vlad starts talking.

"Daniel phoned, you dumped him!" *Oh dear... I hope they have booze at the party.*

"Indeed. A few weeks ago."

"Yes, but only now he thinks you might be serious." Vlad rolls the 'r' through his tongue back and forth as he speaks.

"About time."

"You've got a new boyfriend?"

"No."

And I mean it; during the last hour I have come to believe I am more likely to metamorphose into a kangaroo than get together with Thomas.

"So what is your explanation, then?"

I don't need to give Vlad an explanation, but I know I will. It's just like that with him. And he will say something outrageous or amusing in response.

"I was fed up with him sitting at home doing nothing," I say.

"I understand, he should have found a job. He could have started at Sainsbury's, I have told him many times. So, I tell Daniel to get a job and you take him back."

"No. I will not take him back."

"Up to yourself, but it is the wrong decision. You are old now, and not as beautiful as Lucy. You will not find someone else. You will end up alone with a cat."

His voice is firm, in a way that is typical for Vlad. Why this is not offensive, unsupportive and downright crass I do not know. Maybe because Vlad's simply being too much himself for it to be taken personally. Somehow it makes me smile, gives me a lift.

"I'd prefer a cat to Daniel."

"Okay. I phone him up and tell him there is no chance anymore. He should get over you. I will invite him to Russia to find a girl there."

"Excellent plan."

I'm not sure if I am quite ready for such ruthless practicality. The idea of Daniel going to Russia on a chick hunt gives me a sinking feeling. But I know anything is better than having him sitting at my kitchen table every evening.

Afterwards Thomas looks at me curiously, but I decide against talking about my phone call. We walk uphill through the atmospheric old tenements in the direction of the Grassmarket, a particularly colourful market place in the Old Town.

"I will probably go to Canada… for a postdoc," says Thomas as we pass the Shadow of the Gibbet, a memorial stone in honour of the Covenanters publically executed here in the 17th century.

He had told me during our sunset stroll that he would be finishing his PhD soon. I had refrained from asking what his plans were thereafter. I was fearful that our romance would be interrupted too quickly and our roads would part. That he would not stay in Edinburgh after his PhD was crystal clear. Like me, he wants to become an academic, a professor one day. He will walk on a tightrope from one temporary contract to the next, cajoled along by whatever grant money or contract is available at the time. He will live like a nomad, or perhaps a refugee, until he lands a permanent position. The youthful belief that two PhD students can find two half-decent positions in one town barely lasts a year in the brute environs of the ivory tower. We have outgrown that romantic illusion and know that fulfilling the academic dream means either leaving partners behind, starting with new partners, having one partner settling for a B-job to follow one's academic career or trying to sustain the heartache of long-distance relationships – which pretty much nobody can for the likely number of years. Academic careers and romance are oil and water. How much that might have saddened me two weeks ago… How happy I am tonight to contemplate a period of forced promiscuity…

"Why Canada?" I ask.

"I want to go as far away from Edinburgh as possible."

"Then you should go to New Zealand. Anyway, why as far away as possible?"

"I want to forget about her."

Oh God, he wants to be rescued or he really does think I'm a social worker. I sigh, and let my eyes rest on the dark graphite facade of the Maggie Dickson's Pub, a touristy place named after the legendary woman who mysteriously awoke a few hours after her execution. *I feel a bit dead myself…*

We continue along West Port into Lauriston Street, passing a few strip clubs which – at all times – have clusters of men smoking outside, much more than other pubs; stripping and smoking seem inextricably intertwined.

I point to an apartment building blaring out noise. "That must be it," I say. When the door opens we walk up to the third floor. "Us coming together will create rumours," I say.

Thomas shrugs, indicating that he couldn't give a hoot about potential rumours. *Of course he doesn't care what other people say or think, or feel for that matter; he has made that plain this evening… But I don't want all my chances on the market to be spoiled…*

The hallway is full of familiar faces; all students and postdocs at the University of Edinburgh. I am trying to ignore the curious looks we receive. I do not want to be seen with Thomas who, rather inconveniently, I invited. But I can't say to people, "Actually we're not here together because we are having sex. In fact we're here together precisely because I decided *we are not having sex.*"

"I will look to see if Lucy's here already," I say to him, and slip off.

I walk up and down the flat and finally see Lucy standing at a window with two guys. They are undoubtedly trying to chat her up but it's hard work because Lucy doesn't flirt. She barely talks. Still, as usual they will not be deterred easily, and they probably know that every guy here wants their space. She just stands there passively, as if submitting to a duty, saying nothing. Lucy rarely says much when there are lots of people around, or even when there is more than just her and me at the table. Initially, I assumed she was bored in public, but she isn't. She likes listening to other people. She has opinions, but keeps them to herself. I've never seen her initiating small talk with anyone apart from with me that day a few weeks after I started my PhD. It was a strange conversation, but she opened it. She has never needed to learn to start a conversation. There are always people wanting to talk to her – mostly men but she impresses women too, just as effortlessly.

Lucy greets me with a warm smile and, for a moment, the guys glimpse hope that their encounter with Lucy will be normalised. But she walks away from them towards me. To them, she possibly seems superior and arrogant now, yet she isn't at all.

"Hi, good you came!" she says. "Are you alone?"

"No, the freak is here somewhere," I say, stressing the word "freak."

"OMG, what happened? He got the matrices out, didn't he?"

"Nooo… I wish. His room is covered with pictures of his ex, *everywhere.* Seriously he has more pictures of her hanging around than you would find crosses in a catholic church."

I take the bottle of red wine out of my handbag and two plastic cups that look half-way clean from a table. Lucy whispers: "I heard once that he keeps her underwear as well."

"What?! Why didn't you tell me?!"

I laugh at this bizarre idea, which I can now believe.

"I didn't think it would be true. I thought it was just daft gossip, I didn't want to spoil your fun."

She looks at me as if she terribly regrets not sharing this information earlier.

"Well, *I'm* not sure it's *untrue.*"

"Let's get out of here… and bring the wine."

Lucy gets a 10-pack of Marlboro out of the back pocket of her jeans – our excuse for leaving. I pack the bottle of red wine in my handbag and help myself to a bottle of Rosé as well. We move slowly through the crowded corridor to the front door, then stroll off into the freedom of the lively city streets.

We stop in The Meadows, a large inner-city grass-land with tree-lined tracks. Groups of youths and adults have covered the park with picnic blankets on one of Edinburgh's rare warm summer evenings. We sit on a bench, with the bottles of alcohol and light our cigarettes. It is mere seconds before the first passing guys discover Lucy and stop to chat. We wave them off and continue drinking. With regular interruptions from passing men, we tell each other anecdotes and stories until the early morning. When I get home I sit on the windowsill, reluctantly dialling my answerphone service.

20:17 p.m. Hi, Ka, It's me. Daniel. We really need to talk. I think you made a mistake. After so many years you can't expect passion to stay, of course some boredom slips in. Put it aside, it will not get better with anyone else.

20:45 p.m. It's me again. I found this amazing place in Italy. We can sleep together in an olive orchard. You will love it. I just sent you the link by email. Let me know what you think. *Sounds good… with someone else… Anyway, who'd pay for that trip?*

21:20 p.m. I think you should go to a relationship counsellor. They can confirm that we are a good match… that it won't get any better. If you think it helps, I could join you. *That is the point of relationship counselling, you go together, but let's just save everybody's time.*

23:04 p.m. Please Ka. Phone me back. Where are you?

0:05 a.m. For fuck's sake Ka! Call me as soon as you hear this message, I know you're listening to it! *Oh please move on, not for my sake, I deserve whatever pain I feel, I brought it on, but move on for your own sake, just roll with life's punches and recall we were miserable together…*

I leave my telephone on the windowsill and go to bed. I need sleep. That's all.

Chapter 22

I just manage to roll my body to the other side of the bed and, with eyes barely open, grab in the direction of the godawful noise. It's my old Nokia telephone, sounding super-annoying. The stabbing pain in my head is unbearable. The phone's display reads "7:00 a.m." Autumn light penetrates the bedroom curtains and thence into my brain. Sitting up straight takes at least thirty seconds, lest I empty my guts in my bed. I walk, sort of, to the bathroom where my dehydrated body proves incapable of peeing. I try another orifice; head over the toilet bowl and fingers in my throat. No luck there either. *Why did God not bless me with a body that can handle the lifestyle my heart desires?*

I chuck the last bit of bitter Lidl's coffee in the French press. Let's hope my taste buds are defective so I can drink this sludge. It doesn't take long to find Ibuprofen. I've become a boon to the pharmaceutical industry in recent months; strips of painkillers can be found all around my habitat. I guess a status switch on Facebook, from "relationship" to "single" invariably means a breakfast switch from "nice bread with fresh juice" to "Ibu with coffee."

With one flat tyre, my bike leans sadly against the stair balustrade. It had been flat already yesterday evening when I left the pub with Lucy, so I walked home, but then it slipped my inebriated mind. *Shit, I don't even have tyre levers or glue, or whatever you need to fix it. My next boyfriend will be a dab hand at fixing punctures – it's a sine qua non for sex.*

I opt for a quick and horribly cold shower then make myself go to work. For the first time since I started my PhD I await a bus, on Robertson Avenue, to take me to campus. The rain is drizzling down and I repeatedly

© Springer International Publishing AG 2017
K. Bodewits, *You Must Be Very Intelligent*,
DOI 10.1007/978-3-319-59321-0_22

check the bus schedule hanging on the streetlamp. A handful of people stand on the pavement, hiding under umbrellas. I don't have one – never have had. The concept of umbrellas in windy regions like Edinburgh and the Netherlands has never struck me as tenable. An old, short and thin lady has opted for a long trench coat and a transparent plastic headscarf which slightly crushes her blonde perm curls. She holds a stick instead of a brolly. She has a friendly face with alert eyes and I ask her if I am at the right bus stop?

"What did you say my child? Where do you need to go?" she asks while lifting the plastic headscarf to free one of her ears. She talks with a strong southern English accent – lovely and easily understood speech.

"The university campus, King's Buildings."

"Oh yes my child, it will come. Isn't it dreadful in this weather?"

"It is," I say, not really caring about the rain at all.

At a low point in my life I ponder the question: is everyone here – is anyone here? – as miserable as me going to work? Bang on cue, the friendly woman takes my right hand softly in a way my granny might have, standing in the dwarf door of her little witch house. "May I enquire? What do you do at the university?"

I am probably eight inches taller so she looks up at me. Her skin is wrinkly, but she has a youthful glow, somehow. Her eyes express hope and happiness, like a little girl asking daddy for another pony ride.

"I am doing my PhD there."

"Oh… In what field?"

"Chemistry."

She slaps her hand in front of her mouth, looking ever more surprised.

"Wow, you must be *very* intelligent."

I lift my shoulders, not knowing what to say. I feel quite astonishingly unintelligent this morning. After a short moment of uncomfortable silence the old lady says: "You should be proud."

She gives me a supportive look. And for a moment I feel as if she could be my grandmother watching my graduation ceremony. I am in that delicate, raw, hungover state and I feel my body filling up with emotions, again, the same random emotions I managed to drink away quite well yesterday evening. But these ones are more about my PhD than my relationship, I think.

"Maybe," I say smiling.

"Where are you from?"

"The Netherlands."

"Oh lovely. I've been to Amsterdam a few times… going with the boat from Harwich to Hook of Holland. I was so sea sick… ill over board… years ago when I was still able to party into the small hours."

She laughs, shaking her hips playfully and holding tightly to her stick. I am slightly shocked and ask if she's okay? The hip-shaking soon has everyone at the bus stop staring. A few people laugh but others look worried, like me.

"Oh, come on, I'm not THAT old!" she squeaks loudly, while looking around the full audience, her arms and stick lifted up in mock protest. "I will show you something."

She slowly walks to the metal fence separating the pavement from the private property behind. She hangs her stick carefully on the bars and claps her hands until a few of the bus stop audience softly start clapping the same rhythm, while trying to keep their brollies above their heads. She places her hands just above her hips and starts making large dance steps, lifting her knees high into the air. It looks ropy but she is keeping the rhythm. Every time a foot hits the floor she claps a heel on the pavement twice. It takes only about thirty seconds of warm-up before her hands leave her hips to make full-blown dance moves. From the way she moves one assumes she has full control of her old body and skinny legs but still a few people are poised ready to catch her if she stumbles. She seems extraordinarily happy. And the audience, including myself, is suitably entertained for a few minutes, until the bus arrives.

"That was awesome," I say, handing back her stick.

Her breathing is heavy and she struggles to speak. Nevertheless, she manages a very friendly greeting for the bus driver. The bus is crowded and a young guy stands up to offer her a seat. She takes her plastic headscarf off and blonde curls jump out like spring feathers. She looks slightly younger now, but still she must be at least eighty.

The bus is staying far south of the city centre, driving towards the east. At walking pace we pass the posh, private school, George Watson's College, where parents clog up the road with proud, shiny SUVs. The bus engine makes a worrying sound as we overtake a small group of high school students, all attired in perfectly ironed uniforms.

"If I had a child I would never bring it here," says the old lady, throwing a disapproving look at the SUVs.

"You don't have children?" I ask.

"Oh, no. Children were not for me. I married late and with the wrong husband. Not the type of man you would want to present as a dad to your children."

She makes slapping gestures with her hand. I would like her to tell me more, but it seems inappropriate to ask.

"I married a Scottish man, but he isn't alive anymore. I am free."

"You don't want to go back to England?"

"Oh no, I love Edinburgh. It's a good town, and the people are lovely."

The motor of the bus is working hard, making its way up West Mains Road. I press the stop button. "You need to get out at campus too?" I ask, already knowing the answer.

"Oh no, my child, I will go a few stations further to the Royal Infirmary. I need to visit my cardiologist."

"Your heart seems fine."

She giggles like a schoolgirl and says: "One needs little risks in life."

"True. It was nice to meet you."

"A welcome pleasure."

When the bus stops and we can see the statue of Joseph Black the old lady says: "It is a very good university."

I smile and jump out wishing I was as happy as her.

I enter the lab much later than I normally do. My feet are wet and cold. All my colleagues seem to be delayed; just Babette working at the bench, giving me her customary angry look the moment I enter, and soon emitting harrumphy-grumpy-snorty noises – the usual fare. It sounds so automatic I sometimes wonder if she knows she does it. "Good morning, Babette, a pleasure to see you! What a lovely day." I speak enthusiastically while dropping off my coat. I spray isopropanol over part of the bench to clean it and walk to the hot room to get a few Erlenmeyers containing _E. coli_. I empty the flasks into six centrifuge bottles and use scales to check if they all weigh exactly the same, before placing them in the centrifuge. It's not rocket science; most tasks in the lab merely require patience and minimal concentration.

Babette storms out the moment Lucy comes in, who looks in bad shape.

"Hangover?"

"Yes. But you look much too fit after two bottles of red and a beer."

"I feel miserable. But I met a happy person today – they do exist…"

I start telling Lucy about the old lady at the bus stop but suddenly the lab door swings open at brutal speed; had anyone been behind it they would be seriously hurt. It is Vlad. He pauses dramatically in the doorway, holding a small box of ice and scanning the room furtively. Realising there is no one here except Lucy and me, he steps inside.

"Pleasure to see you both," he says, focusing solely on Lucy.

He is smiling, trying to be charming but actually being cringey.

"Oh Vlad, it is an absolute delight to see you as well," I say.

"Are you married yet, Lucy?"

"No, still waiting for a prince to come along."

"The last one fell off his roan, so she dumped him," I add.

"I'm still available; I could get you a horse," Vlad whispers playfully, knowing he doesn't have a chance with Lucy.

"Great! Make it a Shetlander, it'll fit in my pot."

"You want to cook it?!"

"What else would you do with it?"

"Okay, I better get you something else than a horse. But listen, we need to find you a new husband, Ka... urgently."

Though addressing me, he is still looking at Lucy.

"Yes," I say. "I thought about that chubby Chinese noodle chap in your office."

"Benjamin?"

"If he is called that, yes."

"I think you might be in his league. As you are quite tall he will not crunch you when he lays on top of you." *Don't visualise this... don't, don't...*

"Good point."

"I will tell him to wear a clean T-shirt and I will give him step-by-step instructions on how to start a conversation."

"Sounds perfect. Why are we being honoured with your presence? Did your lab mates kick you out?"

"I don't get kicked out by my lab mates, Ka. They would not dare."

I laugh. He looks harmless, too much like Mr Bean, never a Bruce Willis whatever fantasies he might delude himself with. I cannot imagine anyone ever being afraid of Vlad.

"Anyway, I just wondered if I could have some BL21 competent cells from your magic freezer."

"Sure, anything for a poor guy like you."

"I'm not poor, Ka. I can buy Lucy everything she wants. It is just my supervisor who does not apply for any significant lab funds. Or maybe he does apply, but doesn't get any as he has not published for over twelve years now."

"Twelve years!!?"

Vlad has a different supervisor than Daniel had, but apparently there are high-flyers to be found on the eight floor of the Darwin building.

"Yep. I also don't know what he is doing the whole day. Maybe sitting in his office drinking tea. I was thinking about this: I start to see the upside of a long period of limbo, during which you worry, you fret from one fixed-term contract to the next one before anyone gives you a professorship. All those years of job insecurity, at least they make sure you work to prove you're good! These Darwinian years do indeed separate the wheat from the chaff."

"How poetic!" I think about what he said for a moment. "However, lots of excellent but ground-down wheat falls by the wayside with the chaff," I say.

"Hm. Yes. That is true. But shit happens." *Is there any bottom to Russian fatalism?*

I dig through the −80 degrees freezer to find the right cells, all the while wondering if both the Darwinian and the cuddly paths are wrong and a half-way house is missing from modern academia?

"Maybe there is a better method," I say. "Certainly, blatant laziness should not be permitted while some of us are enslaved!"

"Life is slavery." *Nope, there is no bottom to Russian fatalism...*

"Do you happen to have the ptrc99a plasmid in your lab?" I ask, placing the Eppendorf with BL21 cells in his ice bucket.

"I'm afraid not."

"Could you send an email round the Biology Department asking if any-one else has it?"

"I'd better send you the mail list, and you ask for it yourself. I don't have a good reputation. They probably won't help me."

"Really? A skilled diplomat like you, Vlad?"

"I better leave with the cells before your supervisor comes."

Vlad sounds strangely insecure.

"He won't eat you."

"I'm not sure about that, he looks like he might."

He opens the door and turns around once more. "Come to our lab for lunch, I have some nice cheese and olives in the fridge. You might like it as well, Lucy."

One has to marvel at his persistence in the face of all-out, constant and mocking rejection.

"We have got a lab meeting at 11:30 a.m. I'm afraid we won't make lunch," I say while the door is closing.

"A lab meeting?" Lucy asks, her face looking even paler than when she entered.

"Yep."

"Shit."

All lab meetings are dreadful, every last one of them; a sort of oppression booster lest we forget.

I place the cell debris in the freezer and look at the clock; still some time to kill before the lab meeting. I get the agar plates, which I spread yesterday, out of the stove. I expect to find single colonies of *E. coli* grown on the agar but the plates are empty. I throw them in the bin, then sit in the office resting my wet feet on the dusty heater behind my desk, letting time pass by. *This is not what I ever thought I would want to do en route to becoming a doctor?*

With German punctuality, Logan locks the office door at 11:30 a.m. and we walk together from the lab to the seminar room. "Ready for another morning talking about KBL?" he whispers.

"Oh yes, can't wait to hear the latest! You never know, maybe we will talk about your project instead?"

"I doubt Mark even knows the name of the enzyme I'm working on."

"Probably not."

Indeed Mark does not have a clue what Logan is doing. We all have long recognised that Logan's project, which is industry-funded, is permitted solely because it brings in money – Mark has no interest in it whatsoever. There are several generally unrelated projects running in our lab. The KBL project is Marks' favourite. He is obsessed with it. I think if someone were to ask Mark to rank the projects he supervises, there would be lots of empty space after KBL, though mine and Hanna's would probably be next, followed by Erico's then Quinn's. Lucy's would be fourth or fifth and Logan's would be at the bottom if Mark remembered to add it at all. The injustice of his support, both financial and scientific, angers most of us. It is like having a handful of children being allocated social benefits and only one being fed.

For the first fifteen minutes Babette presents her KBL research with short sentences, hectically waving at different PowerPoint slides. Her voice quickly becomes husky; she is clearly unaccustomed to speaking at length. I ponder what it is like for her family when she goes home for Christmas. Does she get angry when other people sit at the same table to have food? Or what if someone dares to say good morning to her?

After the presentation the meeting goes on for over two hours, exactly the way everyone expected it to; Mark, Barry and Babette talking about KBL, and the rest of us just room-fillers- sort of figurants adorning their higher dialogue.

Every now and then Mark asks us a question, I suppose trying to involve us in the discussion, but he mostly ends up shaking his head – after correctly sensing a profound absence of interest and knowledge in every response. He doesn't understand that we do not share his passion for this one single protein. I guess most of us could have developed some interest by now, but the unfairness of the ludicrous overemphasis kills the sharing spirit.

"Any other points to discuss?" Mark finally snarls.

He looks round the table. We are all thinking the exact same thing: *I will kill anyone who speaks.*

It is the unwritten rule of Lab 262 that these moments are not suck-up opportunities. It is often at this point that people go to war. It is mainly

Babette and Quinn sniping through vituperative gun sights – spraying bullets of derision at each other – and they usually load up and fire a bit even before this starting signal. Now they both let rip big-time. The rest of us pre-emptively wear stony expressions to deflect stray insults until the show is over. I confess to a certain *Schadenfreude* – some days, but not today. It is too warm in the room and the oxygen level is too low. Headache *without* nausea is the best possible state my body can hope to achieve right now.

I sense my mood going from baseline boredom to aggression, my head-ache is throbbing and I know my blood sugar has dropped to the bare minimum. I need food. Thankfully Babbette and Quinn react slower than normal and for a moment it seems like nothing much more will happen. Barry sighs and lets his arms hang next to his chair, clearly waiting for some-one to call an end to this dismal gathering. Hanna looks like she might eat Barry if she doesn't get hold of her daily bagel very soon. Even Lucy's eyes have lost sparkle. Just when Babette opens her mouth and wants to fire, Logan says: "I guess we are all quite hungry and need some food."

Babette moans and sputters, almost soundless.

"I agree," Mark says, ignoring Babette's disgruntled expression which includes the weird exposure of her gums. "Another time Babette, okay?" he adds and smiles at her.

We all stand, lift our notebooks and walk towards the door. "Just one thing, Quinn. Eva needs to move flats this weekend in Stirling. It would be good if you could drive up there on Saturday morning and help her." *I beg your pardon? Why exactly would Quinn help your girlfriend move?*

Quinn lifts his head and pauses for a few seconds before turning around to face Mark. He lets his tongue roll over his lips and then bites it. His eyes are screening the door as if considering his options and looking for a reply. Slowly he turns around and says: "Sure Mark, what time?"

"Nine is early enough. Eva doesn't like getting up early." *Quinn neither, and to be in Stirling at nine means rising about seven in Edinburgh – it's a one-hour drive.*

"No problem."

All of us, except for Mark and Babette, leave the room. A few steps down the corridor Quinn punches the wall. "Fuck!"

"Why did you say yes?" I ask.

"Why?! Because I should have finished my PhD three months ago already! I want to have some chance of getting out here!"

Most of us head to KB House to score a sandwich which failed to sell dur-ing lunchtime; stodge that recharges very minimally. Back in the lab, I still don't feel up to work. Maybe I never will – it's a liberating fantasy… I fetch cigarettes from my desk drawer. Lucy notices and joins me downstairs between the chemical waste containers.

"This colleague of Felix's, Simon, he is really weird," Lucy says pointing at two guys smoking outside of the gate. "Did he tell you he only has one ball?"

"That was indeed the first thing he mentioned. He actually told me a lot about his genitals before I even knew his name. I think I know all measurements in all dimensions, both erect and limp, but not if it's imagined or real."

"Freaking weird."

"Maybe he wanted to raise curiosity? Aren't you interested to know what it is like to have sex with a one-ball men?"

"I don't feel that desire, no. You?"

"No."

We walk back upstairs and start to work.

Feeling defeated, I stand at the bus stop in the evening. Felix, passing the bus stop on his bike, sees me and stops.

"You take the bus?"

"Not normally. I got a flatty yesterday after the pub."

"You need help fixing it?" *I definitely do.*

"Is that not stereotyping me? Do you think I am a woman who does not know her way around a toolbox?"

"Up to you." *Fix it, oh please fix it...*

"Okay, if you don't mind."

"Where do you live?"

"Twenty-six Gorgie Road."

"I'll drop by tomorrow morning before work, around eight."

"Thanks," I say, feeling quite dreamy and looked after.

"You all right?"

"Yeah, just this strange thing. Mark told Quinn to help his girlfriend move flats over the weekend, all the way up in Stirling. Quinn clearly doesn't want to help, but is going anyway, because he is afraid he might otherwise be stuck in his PhD forever. I don't know... it just seems wrong."

"Ka, if your PhD supervisor asks you to paint his house, you don't ask 'why?', you ask 'what colour?'"

"But it's wrong!"

"It is."

Chapter 23

"I can offer you a coffee at the university as thanks," I say to Felix while he fixes the tyre.

"You got your own machine now?"

"No, I've got bags with instant coffee," I lie, presuming he would never opt for that.

He smiles at me with childish purity. "We can get coffee at my office, okay?"

"I was betting on that."

We cycle together to the university. Despite me finding myself quite well-trained after cycling a year over Edinburgh's hills and a few trips to the Scottish Lowlands, it is a challenge to keep up with Felix. We can't really talk during the journey, as the Edinburgh streets hardly allow for one bicycle and thus definitely not for two riding abreast. I am watching the scenery which still, after more than a year, impresses me. In Edinburgh it doesn't matter if it's raining or sunny, the city is always beguiling with its dark Harry Potter charm.

I am totally out of breath when we reach campus.

"You're quite quick on that tank," Felix says.

"Tank?"

"Yeah, that's not a bike you're riding. It's fucking heavy."

I look at my bike made out of steel and totally equipped to travel the world. I never really saw the downsides of it, and could definitely not imagine replacing it with a lightweight road bike.

"I like it."

© Springer International Publishing AG 2017
K. Bodewits, *You Must Be Very Intelligent*,
DOI 10.1007/978-3-319-59321-0_23

We enter the department together and pass Lab 262 first to fetch a coffee mug.

"Nice workspace…" says Felix, looking around at all the mess.

"Lovely, I know."

"Why did you start your PhD here?"

"Good university. Young supervisor."

"Right… carcinogenic solvents and no fume hoods… and don't let the GMO safety officer see you pour genetically modified bugs through a normal sink!"

"We 'disinfect' them beforehand."

"With this pink stuff?" Felix says, barely suppressing a laugh, pointing at the many flasks at the sink.

"Yup."

"You will get into so much trouble if anyone sees this."

"No. Mark is the GMO officer of the chemistry department," I say, giving him a wink.

"Oh Jees."

I take the mug and push Felix out of Lab 262. We walk downstairs to Homer Simpson's palace and Felix opens the door. It is quiet here. "Where is everyone?" I ask.

"They will come. We don't have many early birds here."

He places my mug under the coffee machine and checks if there are enough beans in the machine. We then have to fetch the water tank from eccentric William. We find him in the lab smashing glass valves with a golf club. Felix tells me he wants them replaced and is not happy with the patch-up job that Pipi Longstockings does with glue. As he seems to be very serious about his business, we first let him finish.

"Right, I'd like to see Elizabeth fix those!" he says triumphantly.

"I see," Felix says sombre like a psychiatrist on his daily visit through the ward. "Something completely different, William. Do you happen to have the water tank for the coffee machine in your drawer again?"

"I do."

"Would you mind lending it to us?"

Without saying a word, William walks into the cubicle where Elizabeth is standing. Her eyes follow William and the golf club carefully while he unlocks the drawer. She looks a bit scared, but nevertheless keeps on whistling – the same tune as always. It seems to be a one-song-fits-all tune for happy, sad – and scary – moments.

William hands over the water tank to Felix and utters his eternal refrain: "Please bring it back when you've finished."

"Sure."

We exit immediately, leaving William and Elizabeth alone.

"What is the suicide rate here?" I whisper.

"Why?" Felix asks, probably knowing in which direction I want to bring the conversation.

"This whistling…"

"I know!… It is like nerve pain."

"Do you think William has always been this way? You know, ready for a mental asylum?"

"I honestly don't know," Felix says and we both watch the last drops of coffee falling into the cup. Felix executes the coffee-making process with the same smoothness he deployed replacing the tubes below the column the other day.

"I am not sure it is only the whistling. William has been a very successful PhD student. He published a ridiculous number of papers. But, since he came here things don't seem to work out for him. And even if his research starts to bear fruit it won't bring him as much as he might wish for. He has the wrong surname." Felix pauses. *What is he talking about?* He looks me in the eyes and, for the first time, he does not have a playful expression. He continues in a bizarrely anguished tone: "We are the only natural science group who publishes in alphabetical order… William Rusthaus is pretty much at the end of the line."

"What?!"

"It is a large lab. Homer wants to avoid conflicts."

"But then no one knows anymore who has done what… the order of authorship is the organigram of science…"

"Yup. Homer is a big shot trying to play God. The result is that the wrong people are credited. But, what's worse is that he does not ALWAYS publish in alphabetical order. Occasionally Homer breaks his own rule for whatever kind of reason. Now explain THAT in a job interview."

"A disaster… kind of alphabetical discrimination?"

"If you pay attention, you'll see that a lot of people working here have surnames starting with A or B. Some apparently knew about this rule beforehand. William didn't."

"You start with an H, right?"

"Yes," he says bitterly. He looks disappointed as if his career took a body blow from almighty Fate. "My cards are not stacked well. Maybe I should quickly marry someone whose surname starts with A or B."

"Is that why you fixed my bike?"

"Shit, you uncovered the master plan!" he says, theatrically grabbing my free hand.

"Did you know this beforehand?"

"Nope! Otherwise I would never have come here."

"Why don't you leave?"

"I don't know. Maybe because I want to become a professor one day and therefore I need a successful postdoc. The academic system is not forgiving towards missteps or lost time. I know it sounds naïve, but I hope that in my case Homer will decide against his alphabetic order."

Felix leaves a long rhetorical break. "And, like you Ka, I moved to Edinburgh; the costs attached – and I am not talking only about finances – are high."

I look at him not knowing what to say. It dawns on me that I completely misread this guy. Behind the optimistic, outgoing veneer is a worried person seeing his career slip down the drain because his almighty boss resists tough decisions. After a short silence I try to lighten the mood, "You think Elizabeth is still alive after we left her alone with William and the golf club?"

He laughs and we are back to the people we pretend to be. "I better check on them."

"Thanks for the bike and the coffee."

Walking to the door I pass one of Homers' *Nature* covers, and I wonder if the right person got the glory of being first author out of a list of twelve. *8% chance of fairness, not too bad after all…*

While I am getting things together for today's experiments, Lucy interrupts. She enquires about the previous evening: what happened that I arrived with Felix this morning?… I tell her about my flat tyre and another unexciting evening on the window sill. She is disappointed by the lack of juicy goings-on.

"Well, at least I have something to share. I got a message from someone wanting to date me." She says this as if it is worth saying. It is an almost daily occurrence, hardly breaking news.

"Oh really?" I say, overly surprised. "Who is it this time?"

"You remember the French guy I told you about, the one I met at the party last week – he invited me to the cinema. I know it's wrong…"

"Is he good-looking?"

"Okay-ish. Bit small"

"What does he do?"

"He's an artist."

"Oh God," I sigh. "What kind of artist?"

"A writer."

"What is he writing about?"

"He doesn't know yet."

"He doesn't fucking know?!"

"He is still thinking about it. But in the meantime he wrote me a poem."

"Oh Jesus! A poem? A French Robert Burns… Does he also have lovely, long, triggering sideburns?"

We both laugh. Logan watches us, disapproving as always, but with a smile on his face. He shakes his head, pitying whatever man we're deriding this time.

"Sorry Logan, but poets and gynaecologists are an absolute no-no," I say.

"Agree," chips in Hanna, who just joined us at the bench.

"What is this selection based on?" he asks indignantly.

"Gynaecologists have learned to see the female genitals as a breeding ground for disease, that's not good for anything except their job," I say.

"Exactly!" Hanna says.

"And poets see a double meaning in *everything* and they talk in vague words; both of which are annoying habits. But even worse; happiness is destructive to their art. They need to see the miserable side of life all the bloody time to be able to get something on paper which other people want to read."

Logan stares at me as if he just saw an eland passing outside the window. For him the explanation doesn't make sense. But three women glibly dismissing men out of hand stifle his will to argue.

"What shall I tell him?" Lucy asks.

"That unfortunately you do not have time to listen to his poem as you are busy rehearsing your role as a chicken in a theatre play."

"That is a good reply! I'll keep that one," Hanna says.

I prepare a protein sample to measure at the mass spec service in the Darwin Building and get into a coat. It is only early November, but the snow is falling thick. I hold my hands above the small glass plate to guard the sample against the snow. Carefully I ring the bell next to the locked door of the mass spec service. I've never been here before but have heard that the man running the show is very friendly and competent. A thin, long, grey-haired man opens the door. "What can I do for you, young lady?"

"I would like to make a mass spec of my protein."

I show him the sample in my hands. He takes it and very carefully looks at the glass plate.

"No problem, come in."

He opens the door a bit further so I can pass, introducing himself as Matthew. We walk to the end of the corridor where we enter the small room with a computer and a large MALDI-TOF spectrometer. He hands me a piece of paper to fill out.

"Where do you work?" he asks while I scribble down my details.

"School of Chemistry, doctor McLean."

The friendly man starts laughing, a bit uncomfortably maybe.

"Then you don't have to fill that out."

I look at him, awaiting an explanation. *Surely he doesn't give priority treatment to Mark?! Nobody favours Mark...* He explains, "Your boss doesn't pay anyway."

He places a pencil between his lips, thinking about how to handle the situation. I look at him with pleading eyes. What a disaster it would be if I couldn't use it. I *need* the results. Finally he asks, "You really need it, don't you?"

I nod vigorously. "I do."

"You know what, I will write you down as a member of the Johnson group instead. They pay a fixed fee per year, they won't notice." *Thank you, Gracious Nerd!*

"Does Mark really not pay?"

"He is a pain in the backside. He sends angry emails. That's all."

Entirely inappropriately, I apologise for Mark.

"No worries. It is not your fault."

He enthusiastically starts explaining how the machine works. Most of what he says might as well be in Swahili, but I respect his effort and passion. This mass spec is clearly his baby and he is pleasant company. However, the incomprehensible content of his non-stop talking combined with the claustrophobic setting – a small room without any windows – soon reduces me to stifled yawns. I am not a strong spirit these days and the history of mass spectrometry interests me as much as the history of Boy Band B-sides. After thirty minutes craving for oxygen and a different topic of conversation he finally puts my sample into the machine. We let the mass spec shoot lasers at the target, which looks surreal, almost like being in a computer game. I wonder how much of his fascination with the topic derives from the intellectual thrill and how much from the boyish joy of firing a laser at things. It doesn't take long before a beautiful spectrum appears on the screen. "Hallelujah! That's my protein!"

I save the file and print a hard copy to be sure it won't get lost. It occurs to me that my successful results were begat by way of financial subterfuge. And, miserably, that feels normal enough now.

"Okay. Let me show you a few other tricks that you can do with the machine," Matthew says, eagerly moving the mouse over the screen to change some settings. *Oh no, not again...*

"I am terribly sorry. I would really love to learn more, but I have to go back to the chemistry building. We have a lab meeting starting in five minutes."

"No worries. I can explain next time. Or you can give me a ring and we can make an appointment for the rest of the explanation." *You are a good guy, but I would need a gallon of wine to make it through the rest of the explanation.*

"Do you happen to have a clarifying text file for dummies you could send me instead? Or a YouTube movie I could watch?"

He thinks for a while, then hurries to a shelf at the back of his office and picks up a book the size of a breeze block. "Here. You can borrow that if you promise to return it. Everything is in there. It is like the mass spec bible." *Whoopee.*

"Oh I think this is not a good idea. Look at the weather outside, it will be destroyed. That would be such a shame."

He opens a drawer and takes a Farmfoods plastic bag out. *You really think I am going to cross campus with a mass spec bible in a frozen food specialist's plastic bag? Bye bye intercourse!* He carefully puts the bag around the book while I clutch for another excuse.

"Wait, wait! I will check in our lab first. We have so many books we might have it ourselves. If not I can still come back."

"I don't think McLean has it."

"Oh, he might, really, he loves mass spec."

The poor man looks crushed. I feel bad. I consider taking the breeze block after all. *Me with a Farmfoods plastic bag that screams poverty... No, sorry, I'm not overly vain but I do have limits... And I feel terribly unattractive already...*

Empty-handed I leave.

I walk back to the lab and ask Lucy, "Did you send it?"

"Yes, he already replied. He is keen to see the chicken play."

We both laugh. Logan looks on despairingly.

Late afternoon Mark walks into the lab. He has my first year report in his hands, which I gave him months earlier, just before breaking up with Daniel.

"Good stuff," he says, laying the report in front of me.

I don't say anything. I wouldn't know what to say. Actually I never expected he would bring the mandatory report back up again.

"You will get some publications out of your PhD, I'm sure," he adds.

He looks enthusiastic, the same way he looked at me at the start of my PhD, but I haven't seen him like this for months. I remain silent, unsure what to think. "You can continue the LpxC project if you wish."

"I would like to," I mumble.

I really do want to continue that project, as it was my own idea – it's my baby; the research for which I risked using up all the valuable antibiotics.

Mark follows up with a crazy number of new things he wants me to do. He really seems to think that I am a super-dextrous octopus never in need of sleep. But at least he is friendly for a change. After listening to him for about a quarter of an hour I realise he has something else on his mind, which he only now conveys because he wants to leave the lab, "Oh, Karin, your first year exam is next week, on Wednesday. It will be James, Prof. Gilton and myself taking it."

For a moment I freeze. I knew that this exam would come sometime, in fact we are months late with it, but I'm not ready. I haven't prepared or studied or spared it a thought. I spend endless hours repeating hopeless experiments, trying to make the plasmids I still don't have. And I celebrate my single status with buckets of booze. And now he gives me a week's warning. It is understood that to protest or criticise is a declaration of long-term hostilities with the boss. Silence seems the best – and only – reply. He smiles, insofar as he ever smiles, sort of de-grimaces anyway, and saunters off.

Chapter 24

As Prof. Gilton draws the next chemical formula on the piece of paper in front of me, I feel my hip bones hurting from sitting on this hard orange plastic chair for too long. My exam nerves faded and now aggression is creeping in; I want to crumple the paper up and stuff it down Prof. Gilton's shirt- this man just in front of me who is rumoured to be a notorious alcoholic. I don't know if it's true, gossip spreads like germs at work, and he looks quite sober to me now. For sure Prof. Gilton is nourished by his own mysterious reputation. Why did he give his lab to Mark so long before retirement? And why does he not supervise PhD students anymore? Most days he sits in his office, but apart from reading PhD theses which Mark refuses to touch, and puffing away like a steam machine despite the smoking ban within the building, one can only speculate what he gets up to in there.

"Explain how that reaction works," Prof. Gilton says, pointing at the piece of paper.

I find myself leaning back in the chair, a position that does not suggest much enthusiasm for continuing to participate in this exam. I press my lips together. I know my eyes are shooting fire, but I can't stop them. Just at the moment I feel that I'm losing control over my own body, Mark speaks up, for the first time during the exam.

"It is enough now," he says with a firm voice.

He must have read my thoughts, or at least felt that I am boiling of anger. His interference was necessary to keep the peace. *Thank God. Or thank Mark.*

Prof. Gilton looks surprised as Mark, being my first supervisor is not supposed to interfere with the process. He was keen on drilling me for a bit longer. I have never previously had any contact with Prof. Gilton and,

© Springer International Publishing AG 2017
K. Bodewits, *You Must Be* Very *Intelligent*,
DOI 10.1007/978-3-319-59321-0_24

rumours of alcoholism aside, I have heard only kind things about him. But during the last hour or two I have come to fear him. There is a long silence in the room as if Prof. Gilton is thinking about whether to accept Marks' interference or not.

"Really, it's enough," Mark repeats.

"I agree," James says after a while. Throughout the second half of the exam he watched me with pity, perhaps knowing that the last remnants of my confidence, after a year in Edinburgh, were being pummelled. The questions and pressure from Prof. Gilton destroyed me. He had zoned in on reaction mechanism like a vicious cat pawing a baby mouse. At first I tried to answer but that elicited only rueful shakes of the head. I had hoped we would move on to questions that I actually studied for, but we didn't... only reaction mechanisms. *Is that what I came to Edinburgh for? Did I dodge the world of real work to devote my life to science only to discover that I am too stupid for a PhD? This whole PhD thing feels like one big misprint in my life story...*

"Okay, then we stop here," Prof. Gilton finally says.

He starts to pile up my first year report and the sheets he used to draw chemical formulas and reaction mechanisms.

Prof. Gilton speaks sternly: "As explained before, you can leave the room now, and we will discuss how you did in your exam. Mark will fetch you from the lab when we are finished."

With shaking hands I pack my stuff together and drag myself wearily out of the room which I was so keen to leave. After closing the door behind me, I am not sure what to do. I am panicking. They told me I had to go back to the lab, but all I want is to push over the statue of Joseph Black and leave this department behind me forever. In his humiliating way, Prof. Gilton had given me a message: I wasn't worth a PhD.

Exhausted, worried and angry I am sitting in my office chair, closely observing the all-too-familiar items on my desk. I am looking around the office to see what I should take home when they tell me I have failed. This is the moment they decide if I can continue my PhD or would have to leave now with the consolation prize of an MPhil instead. As I already have an MSc, the repatriation bonus might as well be a paperclip. *Will they really make me fail? They kind of have to. What am I going to do with my life? Who still wants me? Maybe I could become a freezer stockist at Farmfoods, or work in a call centre – if I could explain that I didn't get kicked out for disciplinary measures... This is all just so embarrassing...*

"How was it?" Lucy asks

"Humiliating," I say fully aware that everyone else in the room is listening to us.

"Really? But Mark complimented your report!"

"You think he read it?"

"Nooo…"

"It was horrible," I say. "James asked me about the biology part of my project. It was nice, I could pretty much answer all his questions. But then Prof. Gilton started drilling me on chemistry. He asked me the same questions over and over again. He wanted me to explain how electrons jump in the substrate during enzyme catalysis and how different amino acids in the protein help to catalyse the reaction. I don't know these things off by heart. I never learned the basics. But he asked me hundreds of questions of the same type. He didn't ask anything else, just that, for God knows how long. Mark finally interrupted him."

"They do that sometimes, it is just to make you work harder on the theory for your final defence," Hanna says.

I turn my chair to face Hanna.

"Really?"

I don't remember hearing about anyone having such an experience. *Was it all just a threat?*

"Yes, both Erico and Quinn had very difficult first year exams." Hanna looks at Quinn, who is occupied with a football gambling game on one of the computers. "Isn't it, Quinn?" she pushes him for an answer.

"I was grilled, and it wasn't fun. I don't think I could answer any question they asked me," Quinn says. *You are such a dick not telling me beforehand.*

I feel my worry yield a little and I get some hope – hope that I might be allowed to stay in this nightmare job. It's not Stockholm Syndrome that keeps us here, at least not yet – it's the feeling that escaping this nightmare without a PhD could lead to a life-long nightmare.

"Why did you not tell me beforehand?" I ask.

"What's the point of knowing it? If they want to give you a hard time they will do it anyway. They always find something you don't know. Petra from the Johnson group was asked to sum up all the wave-lengths of the different colours. Do you know them?"

"I learned it at school, but don't know it anymore."

"Exactly, so there's no point."

Quinn directs himself to the screen again, moving some football figures from left to right.

"Did you pass at least?" Lucy asks.

She is worried to lose her best, and maybe only, Edinburgh friend.

"I don't know yet. They are discussing it."

"You will pass," Lucy says.

"You're just saying that to console me, because you feel sorry for me."

Lucy makes no attempt to persuade me otherwise, which is a little dispiriting.

"True," she says after a while. "And quite honestly there is an egoistic side to you passing as well. You still need to enlighten me about the nasogenital correlation theory."

"I didn't think for a sec that you wanted to tap my scientific know-how…"

"What theory?" says Erico, who just entered.

"Lucy wants to know if there is a positive correlation between nose and genital size, and believes I am in a position to find out."

"No way! You are not dating Marius from the bagpipe dude group, are you?" Hanna asks in a shrill voice.

"I wouldn't call it dating. I just kissed him in the pub last week."

"How do you kiss someone with such a large nose?" Quinn asks.

"It's not *that* large."

"It is!" Erico, Hanna, Lucy and Quinn say in unison.

"Otherwise we would not instantly know it is him," Hanna adds.

"Anyway, Ka. Maybe I could be in your control group…" says Erico.

"Sure. Just pull your pants down, stand in front of the autoclave and Lucy will measure it."

I will miss sexual banter. That is what I will lose when I fail, the chance to make and hear cheap jokes.

I turn my chair around and gaze at a scientific paper with empty eyes, the sort of thing I will not be finishing…

Mark finally enters and tells me to come with him. He has a neutral expression on his face, which is uncommon for Mark who normally looks either happy or very annoyed. I can't tell what it means and I don't dare ask. A new wave of panic overtakes me. I feel my eyes getting watery, but manage to collect myself before we re-enter the exam room. It is a gloomy room. The large evergreen fir tree in front of the only window and the relentless rainfall is blocking light. There are two squirrels playing cat and mouse in the tree, undeterred by the rain, slightly lightening up the scene. Prof. Gilton has filled the room with a fresh layer of cigar smell. He and James are talking to each other but they fall silent as soon as Mark and I sit down. After what Hanna and Quinn said, rationally I know there might still be hope, but I feel a strange cocktail of fear and other emotions. The silence is longer than necessary. *Do they get off on every second they leave me clueless?* All of a sudden I feel the nerves leaving my body, making space for a feeling of being in some sort of silly soap opera. *It is just an act, nothing more…*

How much it would have helped me to have this feeling during my exam…
Thanks Quinn!

Prof. Gilton finally breaks the silence. "Congratulations, you passed."

They all look at me, waiting for my reaction. How much I would like to ask him why on Earth he kept on asking me the same questions over and over again. Or why did they want to make me feel so stupid and useless and scared? But instead I simply nod.

"Your first year report is very good, your biology knowledge is good as well, but as you want to graduate as a PhD in chemistry you really need to work on that," he adds.

It doesn't matter anymore what they say. All that matters to me is that I passed. That there is still a chance that one day I will be a professor and I might do something more gratifying with my life than stock Farmfoods shelves. They fill in some admin forms which we all have to sign. Prof. Gilton finishes writing first and looks up at me. "I am convinced you can publish some good work together with Mark," he says. *I would stand a far better chance of doing so without Mark – everybody knows that. Don't they?*

"Pub tonight?" Lucy asks while we are outside smoking.

"I might have to collect some data for your theory first."

I press the send button – a message to Marius.

"Good plan."

"I'll give you a ring later. If it is really shit we can still go for a drink."

Marius has replied by the time we are back upstairs. "He is coming for dinner but he intends to bring candles…" I say to Lucy.

"Oh no!"

Leave the candles at home or wherever you store them and just come over, I reply

"What are you cooking for him?"

"I will buy some bread rolls. We just need the data point, right? It should not become too cosy for him and he should definitely not get the impression I can cook."

I pass Sainsbury's on my way home and buy a couple of dry-looking bread rolls, butter and cheese, a pack of Marlboro and two bottles of wine for the evening. A few hours later I am sitting on the window sill writing Lucy a one-word text message: *Positive.*

I knew it! Call? She replies.

I can't. He is still here… sleeping. You get the next data point!

Was it that bad?

No, was awesome. I don't want you to miss out on the experience.

Hahaha one is enough. His nose is large enough to be foolproof. Talk tomorrow.

I pour myself a glass of wine, smoke a cigarette and think about what happened today… the memory of the exam. *Besides just having proved Lucy's nasogenital correlation theory, will I ever get the confidence back that I could be a good scientist?*

The memory of that exam will plague me like toothache for the rest of my PhD.

Chapter 25

"How was Christmas?" I ask Lucy, shoving my large backpack to the wall.

"Quiet. Bit boring actually."

I flop onto the single uncomfortable sofa in the living room she shares with her flatmate, feeling similarly unexcited about Christmas – just glad it's all over.

We had both been home for a week, and had agreed to fly back to Edinburgh to celebrate New Year's Eve together. I landed ninety minutes ago and went straight from the airport to her flat to help prepare our New Year's Eve dinner party.

It wasn't supposed to be a party; Lucy and I had presumed it would be just the two of us. But some PhD peers learned about our plans and asked to join us. PhDs tend to be something of a lost tribe at big "real world" occasions. For some, flying home over Christmas is simply unaffordable. For others, like Lucy and me, spending more than a few days in the family home, in the small villages we grew up, is unimaginable. What are we to do in a place where we might know each and every person but are now socially bound to none? It's good to be back home though – occasionally. I enjoy standing in front of the skylight in the room that once upon a time had been my bedroom, and thinking back to childhood. My parents' house is part of the last row in the village and looks out over long stretches of flat fields. In winter you can see all the way to the tree line of the next village, but in summer fast-growing sweet corn blocks the view. As children we didn't have many rules to obey, but as soon the corn had been seeded, the fields were forbidden terrain. Especially as the corn grew higher, it became more and more difficult and sometimes impossible to resist the

© Springer International Publishing AG 2017
K. Bodewits, *You Must Be* Very *Intelligent*,
DOI 10.1007/978-3-319-59321-0_25

temptation of running through, even going cross-country and hopefully reaching the sand-mining lake two miles North-West. We would creep across the thin wooden plank over the ditch which separated our garden from the field, and then sneak between the green stalks so my mum wouldn't see us. After our adventures we denied all charges and were always rumbled by the smell and pieces of corncob in our clothes.

The anchor of childhood nostalgia is fun for a few days. It's replenishing and great for perspective on where you're at now but, afterwards, miles of flat fields are quite breathtakingly dull. The village is located twenty miles from the city where I studied, so unless you are a super-fit cycling fanatic or determined not to drink, the city's social life might as well be a thousand miles away. Lucy grew up in a similar setting. Most of her and my friends have moved away, often living all over the world. Randomly settling in any old corner of the globe is a thoroughly modern habit, especially in the natural sciences; it's romantic and exciting on the one hand, but bond-breaking and often lonely on the other.

"Same here, Christmas was a bit boring," I confirm.

"The thing is," says Lucy. "My brother brought a girlfriend home. She seems nice enough but we don't share any common language. She speaks Korean and Swedish."

"Your parents speak Swedish?"

"Nope, only my brother."

"I wish I couldn't understand the moron my sister is dating. I dislike that guy so much it hurts. He is devoid of human empathy and looks like a troll. I kept hoping she would do the classic Christmas breakup thing, but she didn't."

My heartfelt disdain makes Lucy smile.

"So, just a language problem with your new sister-in-law?" I ask, with the suspicion that there is something more personal afoot.

"No, that wasn't the worst. Having – for the first time – a potential daughter-in-law visiting turned my mum into an ecstatic freak. She was super-nervous and over-excited like a tot with a new doll. I didn't recognise my own mum. It was surreal. And even worse; the whole family has been joking about me staying single for the rest of my life. Otherwise why did I never bring a boyfriend home?"

I knew Lucy faced this pressure from her family. After all, she was going to finish her PhD soon and wasn't getting any younger, in her mother's words – and in Edinburgh Vlad was on hand to drive the point home. Her mum and aunties had retired, her cousins were all married and seemed to produce babies like tennis ball machines; everyone desperately wanted her mother to become part of the granny club. I guess Lucy is supposed to pop a kid and fulfill the commonplace desire of sixty-plus women yearning for that new role and purpose in life.

"Just ignore those stupid comments," I say, helping Lucy pull two small tables together to host our guests.

"I tried, but on Boxing Day I subscribed to a dating site."

She is clearly awaiting a reaction.

"You did WHAT?! Why on earth would you subscribe to A DATING SITE?! You can pick any guy you want! They are all begging to go out with you! Men cannot resist you!"

"Most guys I meet and who propose dating me are from the Chemistry Department. And I don't know if you've noticed but they are all freaking weird."

"A few of them are quite hot."

"Hm. I want to see how I get on outside of academia, maybe I want to be out of academia, rescued from academia, I don't know."

"So did you find someone to date? Of course you did – hundreds, no problem…"

"I am chatting with this guy called Luke. Works for the Bank of Scotland."

"A banker!? You might as well take Louis back," I say, referring to a good-looking chap from the Chemistry Department she dated briefly before Christmas.

He had asked Lucy to go on a date. As Christmas was approaching and Lucy did not want to disappoint her mum another year, he was in luck; Lucy was feeling duty-bound and joined him for dinner.

"This is Louis," she had said in front of KB House when they returned to campus after a stroll over the wonderfully quiet and romantic hill behind the university, offering a lovely view over Edinburgh's north side and towards the Pentland Hills in the west.

"Hi, I'm Karin," I said, shaking a weak and clammy hand.

"Hi, I've heard a lot about you,"

His voice was as feeble as his handshake, and as sexy. *OMG what is she doing with him? He is so NOT manly…*

"Oh, did you? Did Lucy tell you I'm not a big fan of Italians?"

Louis looked surprised, not really knowing what to say.

"I know you are Italian and it's not your fault, but as a kid I almost choked on a pizza. I can't do anything about it but since then I can't stand Italian men."

Lucy rolled her eyes. Without words, she was saying: "I know what you're saying: he is not right for me."

"I'm sorry to hear that," said Louis, again in a tone so soft I wanted to roughen his vocal chords with a gallon of whisky.

He isn't sure if I am making this up, which is just plain thick, and it eggs me on, of course. "Don't worry, it's getting better as time passes. I think in a few years I really will be fine, might even go to Venice, though I've heard it's a pigeon paradise with nothing romantic about it."

I raise my eyebrows enquiringly. "I quite like Venice," he responds witlessly, unmanly...

"Really? You should take Lucy there."

"I could do," he says, looking at Lucy like a little pup begging for a bone.

Lucy's eyes shoot daggers at me. I know too well that Lucy would hate going for a romantic weekend away in Venice with any guy. That kind of packaged romance makes her cringe.

"We need to go to the lab now, see you later, Louis," Lucy declared, firmly pushing me towards the chemistry building.

As soon as we were through the door, Lucy looked at me firmly and said: "I know."

"How often have you seen him?"

"About four times."

"And if I remember correctly, you were at his place, in his room, yesterday evening watching a movie?"

While we talked, I carefully took twenty-four Eppendorfs out of the table top centrifuge and placed them in a colourful rack on the bench

"Yes."

"But you did not have sex with him! Right?"

"Hell, no! He wanted to. But as soon as he came close to me I went home instead. I really don't want to have intercourse with this dude."

Phew!

"Then dump him!?"

"I know I should dump him. But I wanted to tell my parents I have a boyfriend. They are starting to worry I'm not normal, that I'm getting too old."

"You just turned thirty, you're not old!"

Lucy didn't look convinced.

"Thirty," she repeated, sinking into a whisper. "No partner, no permanent residence, no clue where my life is leading..."

"It'll be fine."

I said those words in the same voice my mum used them on the telephone after my first day at the University of Edinburgh, when I told her that there is no desk and no computer for me in the lab. They are the same meaningless words I said to Hanna before her presentation at a conference in Venice when I sensed she was nervous. Truth is, more often than not when we say those words we have no clue if things will be "fine." And every time I say them like that I feel like a liar. I quickly add, "I am totally with them about you not being normal, though."

Lucy smiled. "What's normal anyway?"

We place a few chairs around the table and bundle all the mess out of sight. "Tell me about your Christmas," she says, which invites me to confirm that I am as lost in life as she is.

"Nothing special. The first evening I stayed with my sister in Amsterdam, and we had drinks with people from her law firm. It was sort of cool but it showed me how weird I've become in the last sixteen months. I really struggled to converse with the people there. It's like I haven't much to talk about, other than my research. And they are lawyers, they know little about science. Plus, it's weird, like I am just floating, I feel like I haven't been corrected for any behaviour since arriving here. No one has tried to place me in a box, which is good, but nobody has offered me any motive to work on my social skills either... I only hang around with chemists and a few biologists; and they don't care much if I'm a bit off... Mark doesn't seem to care what kind of people we are and how we behave towards our colleagues in the lab. He loves Babette and, rationally, I know that is only because she works on a project he actually likes, but it feels so wrong that she is favoured while she is a monster to everyone else... There's no incentive for anyone in the lab to actually be a nice colleague; Mark doesn't set an example to follow, and success does not depend on behaving decently... Maybe I have been sucked up into the Ivory Tower by my research and by other scientists, and I have pretty much forgotten about the values and unwritten rules of the real world outside... Maybe this is similar to a language you once learned but are now surprised to discover you can no longer speak because you haven't spoken it for so long. We are living in a bubble, the science bubble. And if this is indeed the fabled Ivory Tower then I feel like deflated Dorothy seeing behind the curtain in the Wizard of Oz..."

My voice tails off lest I lapse into ranting about the limbo-land inhabited by all these lost PhD students – I'm not special, my circumstances are not even worth bemoaning.

"You are so right! I also became super weird. I have the feeling I become weirder with every week I spend in academia. I think the only 'normal' person in our lab is Hanna."

"Yep..."

My phone is ringing and I fetch it from my coat.

"Hello?"

"Hi Ka, how are you? Good trip home?" Felix says on other side of the line.

"Fine, fine. Are you not supposed to be in Denmark?"

"I didn't go in the end," he sighs.

"What happened?"

"Oh you won't believe it. It is such a fucked-up situation! Your boss – of all groups in the world – just published an article on how to produce a natural product from living cells – the very same that I'm supposed to synthesise chemically."

"I didn't know Mark was publishing?" I say, sarcastically.

"Well, your dear boss isn't publishing much but he published this article which makes my research proposal redundant. So two days before I planned to fly home, I wrote to my funding body and told them that, given this new development, I think my project should be changed. They agreed that my project doesn't make sense any more, but said they gave the funding for the *project*, not the *scientist*. Meaning that if the project is stopped, I need to pay back all the salary I have so far received. You get the picture?"

"But that's months!"

"Yep."

"I'm sorry, that sounds ridiculous. You can't know beforehand that some-one from a different field will be publishing something that makes your research redundant. It was published after you got the grant!"

"I know, it's tedious, and unfair. They stopped my salary and now they want to decide if I have to pay them back for the previous months. They might give me a second chance after being a bad boy, but I have to deliver a new research proposal to see."

"You didn't get paid in December at all?"

"Nope!"

"Wow, you are being white-washed out!"

"Green washing would fit better as your lab is producing the product biologically."

"Nerd… What does Homer say?"

"That I am an absolute idiot for telling the truth to my funding body and that I should have just lied about the project I'm working on. You know, work on the redundant project here and there to be able to write some rub-bish final report, but spend most of my working time on something else."

"Constructive."

"Anyway, I'm in Edinburgh now and after spending Christmas in the office I wouldn't mind spending New Year with you guys."

"Come over. I'm at Lucy's. Others should be coming soon."

"Cool."

"Felix?"

"Yes."

"I hope William left you the water tank over Christmas?"

"He did!"

Less than an hour later we sit with six people from five different nations around Lucy's makeshift dinner table. We store all the lovely dishes everyone brought from their home countries on the window sill. There are a few bottles of wine on the table, and conversation flows with ease. It is so much easier talking with my PhD peers than with the lawyers and family back home. To varying extents, it's the same for everyone round the table; we're bonded by our similar life situation and concerns which might seem peculiar, horribly stressful and a bit daft to most people. We have all come to understand that our academic success strongly depends upon the resources we can access, the amount of money being spent on the specific project and publication; and towering over these is our boss. Our intelligence, work rate and burning desire to make a difference to the stock knowledge of mankind are all but trifles. My mum had looked surprised when I said that I was nowhere near developing a new antibiotic and my dad had looked downright disappointed. They didn't understand and I didn't feel I could explain. It was as if I had drifted away from the world outside of academia – too far away to communicate my slightly absurd reality.

After leisurely finishing the food, Maggie and Joanna go home in order to be fresh for the lab tomorrow. The rest of us pack the leftover wine and head to Princes Street to see the magnificent firework display emanating from Edinburgh Castle at midnight. The Scots are BIG on New Year's Eve, better known as Hogmanay.

After all the starry explosions and dreamy escapism of the whish-bangs Abel, the cute English guy from the Johnson group, makes a suggestion- "We could go to my place, my flatmates are having a party tonight… I truly don't know what to expect though. They seem a bit odd."

Abel has beautiful eyes, perfectly groomed black hair and a sporty body packed into a classy outfit. Also, he talks with an accent I absolutely love like music. But, through my discerning eyes, he is somehow too perfect to be manly.

"I wouldn't mind checking out what's going on there anyway," he adds, looking rather worried.

"I'd love to but I really need to go home so I can work on this research proposal," says a dejected Felix, who has oddly deep wrinkles under his eyes, from fatigue and sorrow.

Lucy and I follow Abel through the Meadows to the south side of town. It's a long walk but it's a lively scene. I notice Abel getting increasingly nervous the closer we get to his flat. Several times he mentions that his flat-mates are a bit off, and he has no clue what kind of party to expect. He only moved in two weeks ago when his previous landlord decided he wanted to use the flat for another purpose.

We finally pause in front of a building that looks to be mostly inhabited by students. There are numerous bikes out front and a mix of loud music blares from several windows. He opens the front door and leads us up to the second floor, looking shocked at the piles of clothes and toilet paper decorating the staircase. There are shirts, socks, underpants and trousers hanging sadly on the banisters, as if someone has thrown them down from the top floor. The toilet paper looks even more desultory as the product of lowbrow hi-jinks. It is not amusing, just weird. Abel raises his brows and mumbles something I don't hear.

The door to the flat is open and there are people sitting on the floor everywhere. Some are smoking peacefully, some are just sleeping, some are half undressed and all seem to be on a totally different planet. None of us says a word. *Have I been downgraded to a bad Trainspotting film set or am I still a PhD student? Are the two worlds so close these days?...*

We try to avoid stepping on bodies as we follow Abel to his room. The door is ajar and the handle is broken. "They broke into my room!" he whispers, carefully opening the door.

The room looks tidy and a guy with long greasy hair, wearing open sandals in the middle of winter, is sitting on the bedside. His clothes are very old but they look clean enough to suggest he is not actually homeless. "Fuck off and shut the door," he says, gesturing in a manner that manages to be both dismissive and aggressive.

"I'm sorry to disturb you Jesus, but this happens to be *my* room," says Abel.

"There's someone in your bed," I say, pointing at the big bunch of red curls extending from the sheets.

"She must be huge, or maybe they are two," Lucy whispers.

"A whale," I hypothesise.

"What's that lady doing in my bed?" Abel shrieks, before carefully lifting the sheets off the creature.

She looks like a Disney figure. She is wearing stockings and a cartoonish red dress which I doubt you can purchase in an ordinary clothes shop, set off with a Minnie Mouse hairband – a sort of crown of self-ridicule.

"She just needs a rest, buddy," says Jesus.

"Did you have… I mean… did you penetrate her in my bed?"

I guess Abel's Southern English accent doesn't allow for a less selective choice of words, but it seems strangely formal amid the scenery.

"No!" Jesus says, shocked, and evidencing disgust at the creature sleeping next to him.

Abel looks suspicious. "Would you mind taking her somewhere else?"

"She's too heavy, mate!"

Abel looks disturbed. He walks to his desk and lifts an empty bottle of whiskey. He reads the label carefully, looks at Lucy and me, and says: "They drank Steve's special whiskey."

"Who is Steve?" I enquire.

"My only sane flatmate, a maths PhD."

"And he is sane?" Lucy asks in disbelief.

"Isn't he here?" I ask.

"No, he's visiting his parents in the US. He got this bottle from this granddad for his graduation. He is not going to be pleased."

"See the positive; he still has the bottle," I say.

"Yes, he could put water in it, or cold tea to look like whiskey…" Lucy adds.

The numerous bookshelves above and around the desk are filled with novels, chemistry and biology books, CDs and a few pictures of Abel with his siblings and parents. One of them is of Abel holding flowers and a master's certificate in front of prestigious University College London.

Abel, who normally doesn't smoke, asks me for a cigarette with a shaking voice, and adds, "This is really quite fucked up!"

The people still able to move have now decided Abel's room is the place to hang out. More and more filter in. A tall guy, with considerably less metal adorning his skin than the party average, walks up to Lucy and enquires, "You got acid by any chance?"

Lucy takes a step back. "What's that you are looking for?"

"Acid."

Lucy looks confused. "We have hydrochloric acid in the lab, not here."

The guy stares at her uncomprehendingly, but he quickly surmises that Lucy is probably not the sort to be carrying a stash of illegal drugs.

I ask Abel, "Do you mind if I walk around a bit? I'm curious to see what else is going on in your flat."

"Me too!" says Lucy, excited.

"Go for it."

We step over the people in the hall and enter a room that is probably the living room. It is tiny, crowded with about ten people in it. Most are sleeping or at least dozing away. There's a topless girl sitting on the sofa with a guy next to her touching and licking her boobs in a highly unmotivated fashion. She lets her tongue circulate around her lips. Despite these sexual motions between them, it is so perfunctory it reminds me more of a battery-powered Christmas toy moving mindlessly in a shop window than of anything that might lead to sex. It's a drug-driven default setting, repeating the same moves over and over again.

"You think they are always like this?" Lucy asks.

"I guess they are not on GHB all the time, no."

"I've never seen anything like this."

"It's a bad movie."

"Poor Abel."

"The monster in his bed…"

"And living here all year round…"

"They've got a karaoke machine!" I say, pointing at the box next to the sofa.

"Cool. Let's sing!"

Lucy sounds very excited, as if her dream for New Year was to perform karaoke songs to a zonked-out shower of faintly sinister stoners.

We turn off the music coming out of the stereo and start singing our lungs out.

Lucy and I are both far from talented, and our voices are not enriched by the amplification of a cheap karaoke set. The faces of some of the people lying around express physical pain when our singing fills the room, but they are bereft of the ability to move. It's incredibly loud.

"It's good to have such a grateful audience," I say to Lucy.

"Yes, it's fantastic."

Abel joins us. It turns out that he too wouldn't survive ten seconds at a provincial audition for Britain's Got Talent. Does he want to participate in the party or take his musical revenge?

We sing till sunrise. We feel like we belong here as much as anywhere in the real world…

Chapter 26

I take the SDS-PAGE gel out of the gel chamber, patiently peel it off its glass plate, and lay it in a box with staining solution. In a few minutes the protein fingerprint becomes visible. The big spot of overexpressed protein I am looking for is clearly missing – not a trace of it. I press my lips together and sigh. Just another failed experiment. How much easier and more satisfying it was as an undergraduate to do science, where the practical classes were designed for easy success. We got a manual and just followed the experimental procedure; only people who could not read would fail to yield yoghurt from milk or isolate that one single bug from the university pond. The experiments got more challenging during my master's, but still I was jumping onto running projects, or I was working in industry labs with unlimited funding. This is different, the almost daily pay-offs are long gone; for months I make no progress. Despite sharing a lab with other people, my research is in complete isolation and it makes me feel stupid. How could I possibly design experiments leading to significant discoveries with limited lab funds and equipment? And who is there to help? Where is the brain pool, the people who passed through the muddling stage and could actively help me with the design and interpretation of failed experiments? Obviously I would like to discuss them with my supervisor but that is a no-no. The moment we touch upon something he doesn't know well, or tricky problems are even hinted at, he lets rip with random insults, as if afflicted with the protein chemist's version of Tourette's Syndrome.

I take the pellets of *E.coli* cells I collected yesterday out of the freezer and throw them in the large yellow bio-waste bin, like throwing yet another week of my life away…

© Springer International Publishing AG 2017
K. Bodewits, *You Must Be Very Intelligent*,
DOI 10.1007/978-3-319-59321-0_26

"Morning," Logan says on entering the lab. "Guess what I just saw?"

Logan is still wearing his sporty outdoor coat and has an expression that speaks of both shock and entertainment.

"Bring it on, my day can't get worse."

"Babette and Barry… walking *hand-in-hand* to university."

My mouth falls open. I press my jaw bone back in its normal position. Lucy and Hanna come out of the office and all three of us stare at Logan as if he just told us God has been proven to exist and is in the lab next door knitting socks for Jesus.

"O.M.G." Lucy whispers.

"What?" Hanna asks, almost shouting.

"This is why she has been smiling occasionally lately, and is sometimes even late!" I say.

Babette really had been smiling of late, which is something she traditionally only did for one bizarre reason – Mark entering the room. She still never talked to us, but she was running around with bouncy energy. She evidently had something going on but we didn't know who or what it was – till now.

"Poor Barry, it didn't even seem to have an effect on his depression," Lucy adds.

"Maybe he likes to have Babette sitting on top of him like a werewolf crying for her prey," I say.

Logan looks disturbed, makes the stop sign with his hands, shakes his head, but smiles. "Getting a red steak thrown on his plate for breakfast…" I add.

Lucy muses, "She'll probably lock him up in the cold room to die when he dumps her. We've all seen what's happening with Quinn."

"What does he have to do with it?" I ask.

"They had a one night stand after a Christmas party, the year before you arrived," Hanna clarifies.

"Is that why she hates Quinn so much?" I ask.

Lab meeting or not, at least once a week Babette would explode at Quinn, and he would send salvoes straight back at her. Quinn wasn't innocent, he relished provoking Babette.

"Yes. Babette wanted more but Quinn wasn't interested – the neck bites looked like a vampire attacked him," Lucy says. "You got the feeling it was a rough night in all sorts of grim ways…"

"You girls are awful!" Logan says, smiling, and walks to the office.

I quickly add a few microliters of plasmid solution to the Eppendorf with competent *E. coli* cells, heat shock them, and place them back on ice. Lucy waits, holding two cups to be filled with Prof. Homer Simpson's coffee. We walk downstairs and luckily find Felix in his cubicle. Through the glass window we watch William screwing a high drying rack in the middle of Elizabeth-Pipi-Longstocking's bench. There is no way that anyone could still work there. "What is he doing?" Lucy asks.

"Elizabeth installed drying racks for our glasswear on the sides of all the benches in the lab, but she didn't actually ask us if we wanted them. They're on the side of the bench, so no one minds them. However, not so for William. For days, he would screw his rack off and then Elizabeth would stubbornly screw it back on. I guess he's had enough of the on/off game and is trying to make a point..."

"Right..." I say.

Together we walk to the little kitchen in the large office. "You girls coming to James Watson's lecture?" Felix asks, placing the water tank in the machine.

This is *the* James Watson, of course. Eager Felix has a notebook and pen at the ready. He is positively itching to get into that lecture theatre.

"You mean listen to the douche who discovered the structure of DNA?" I say, feigning lack of respect for one of the biggest scientific breakthroughs of the 20th century.

"Yup, that douche," Felix smiles.

We wait until Lucy's cup is filled and, as soon as the noise of the machine indicates it has finished pumping water, William is present and correct: to secure the precious water tank and lock it in his drawer.

"She won't screw that rack on my bench again," he states triumphantly.

"No, do you think she or you or both of you need psychotherapy to sort this out?" Felix asks. "Her whistling might well deepen the therapeutic effect."

The four of us make our way to the largest lecture theatre, which is on the other side of campus. If William's drawer had lacked a lock I seriously doubt he would have attended this starriest of star lectures. I am starting to like William despite, or maybe even because of, his strange traits, but I still feel sorry for people working with him. William seems to respect Felix and they rub along surprisingly well. Lucy and I are now good friends with Felix, and William has somehow sort of loosely grown on us in recent weeks; weirdness, of one sort or another, being the gel that lumps us together.

"I just went through all the pre-exams of the undergrads," says William, who has been teaching at the department for over a year now.

He shakes his head in a manner suggesting the results have not been awe-inspiring.

"And?" Felix asks, smiling as ever.

William sighs deeply, "You shouldn't let monkeys study chemistry, they just don't get the point."

As we walk out of the door, it seems like the whole chemistry building is emptying at once. There are streams of people on the paths outside, all with the same destination. They could have combined the James Watson lecture with a fire alarm exercise; nobody is going to miss this.

It is overcast, rain looks imminent. On my way to see one of the biggest science stars in the world, I watch a few dozen rabbits jumping around the grass looking for food. There is a plague of them on campus. I don't suppose it matters because nobody is trying to control it. There is a big, fat red cat fighting a lonely war against the rabbit plague. Occasionally he catches a fluffy baby bunny whereupon he will take three days to actually eat it all. The fat cat drags the little thing to a bush and lays with it, or rather pieces of it by the second day, until it's all finished. We've all watched this elongated process in horrid fascination. Fatso's favourite spot in the bush is just before the KB House canteen. However, this dismal creature is not around to blight this exciting day.

"Apparently Watson was supposed to come two years ago, but the University of Edinburgh withdrew the invitation," says William.

"Why?" Lucy asks.

"In one of his speeches just before he was to come here, Watson claimed that Africans are less intelligent than white people, and black people are basically genetically inferior," William says.

"Ah, a highly educated racist," I say.

"Yes."

"But now, a few years later, he is not a racist anymore? Or, what?" I ask.

"He apologised."

"Did he apologise for stealing Rosalind Franklin's x-ray structures as well?" I ask.

This is a legendary bone of contention in modern science.

"Have you seen his TED talk?" Lucy asks.

"I did indeed," says Felix. "He clearly admits that the x-ray forming the basis of the structure was made by Rosalind Franklin."

"Maybe Franklin not getting the proper amount of credit stirred up too much debate around the world and he got too scared to *not* mention her?"

"He is still claiming that Rosalind Franklin did not really want to draw the structure because she wasn't really a chemist," says Lucy.

"She was a chemist," I say.

"We will never know the true story, or who really deserves the credit," says Felix. "It never changes; people stealing each other's work, turning science into a soap opera." "But let's see what he says today," William adds.

The lecture theatre is overcrowded. Every last seat is occupied and already people are settling on the aisle stairs between the seats. It's not really a surprise that someone like James Watson attracts such a crowd; arguably, he did nothing less than lay the foundation of modern molecular biology with his discovery. The people look excited. In our lab we've been talking about James Watson gracing our campus for over a week. I too feel excited, a bit like being introduced to the Queen somehow – apparently a racist male queen, I just discovered, which bothers my liberal sensibility, but still I am excited to be here.

We sit down on the congested stairs, behind each other, and await the learned genius. As per every introduction in the scientific world, the speaker gets complimented on how great his or her contribution has been to the world of science. In the event, this introduction is suitably brief; come to think, in a hall full of budding and accomplished scientists, the introduction could have been reduced to just two words, 'James' and 'Watson'.

Watson walks on stage, slowly. He wears a too-large pair of trousers, a shirt and a blazer. His bald head is covered with liver spots and his back is bent. He is old, very old, and his speech is unclear. It takes me a few minutes to get acquainted with his pronunciation or, to be blunt, his lack of it. He mumbles and mumbles unexcitedly and, to my amazement at least, does not talk about the discovery of DNA.

Instead, he talks about all the great scientists he has met during his long life. I follow it for a while, waiting for actual content, but there isn't any. Out of boredom I look at my watch and see that we are only twenty minutes in. This isn't a lecture, or even a speech. It's just a long boast. "I know very important person A, research Institute B was named after him, and BTW I am really bloody important myself, so there!" That was all there was; a long, bland statement of something everyone in the room already knew.

The people around me look as disappointed as myself. My right leg starts to tremble impatiently and I regret that I did not bring something to read or that I can't kill time knitting. Just about anything would be preferable to listening to embarrassing and pointless trumpet blasting. I gaze round the theatre trying to calculate how much money is being wasted just here and now? Thirty minutes in, I check to see if I can escape, but there is no way through this crowd. I resent listening to the self-love of an old man who is wasting so many people's time.

When the lecture finally finishes, people applaud. *Are they applauding themselves for lasting the whole hour?* The applause is of a peculiar polite character, such that might greet an amateur theatre company on the opening night of a dismal show. Of course the moderator says a few nice words before we all stand up, ready to head back to our labs, hoping we never become like James Watson – the opposite of what we hoped on the way here.

"That was really good," a girl declares to someone else in front of me.

"Really?" Felix says from behind her. "You really thought it was good?"

About twenty people turn towards us, some look surprised, some entertained.

"Quite honestly, I found it really shit," says William. "Probably the worst lecture I ever attended."

A few people laugh, others turn away, not wishing to acknowledge the depressing fact.

"He joked in his TED talk that he doesn't really like to talk about his DNA discovery anymore," Felix says, while we cross campus back to the chemistry building.

"Well, that was pretty clear," I say. "Just an old man babbling."

"If you are old and famous you can serve up this shit," William says.

A few other PhD students, walking within listening distance, agree it was a disappointing lecture.

Two of them introduce themselves to each other. Apparently they haven't met before. "Jonas" one says, "Peter" the other says.

"Jonas?" William interrupts. "I thought you were called Billy?"

The tall, blond guy who just introduced himself as Jonas looks at William, surprised. "You are from Sweden, right?" William adds.

"Yes, I am," Jonas says, still looking perplexed about how people came to think he was called Billy.

"So, it's Billy… called after the IKEA wardrobe," says William, maintaining a very serious face.

Felix is indicating without words to Jonas and the others that William is nuts and should be taken with a grain of salt.

All the students are laughing, though somewhat uncomfortably. A long silence follows.

Just before entering the School of Chemistry, William slaps Jonas on the shoulder. "Was just a joke, pal!"

It was blunt stereotyping, ultra weird and profoundly pointless. And yet infinitely more memorable than anything famous James Watson had to say. How utterly dispiriting.

It is early evening as Mark enters the lab. I feel that all my colleagues are evaporating but as I am pouring plates I cannot escape. My muscles are tensing up as he walks towards me, holding a scientific paper in his hand.

"Babette here?" Mark asks, in his default barking tone.

"No, she left."

Mark looks surprised. Babette rarely leaves early.

"Barry?"

"Left as well."

I consider telling him about the new turtle-dove couple we have in the lab and that I suspect Barry is currently being chewed by Babette, but I don't. I can see in his eyes that despite me being second or third choice he cannot resist talking about the paper he holds in his hand. Enthusiastically he starts explaining about the findings of a paper on a random cofactor-dependent enzyme. During his monologue, he shows me tables and figures. He tells me that some of this research can be implemented into the KBL project and so on and on and on... *I know this will end. Just wait patiently. Keep yourself quiet like a dead body and you will be fine.*

Abruptly he changes topics and tells me about all the fantastic things Brian discovered lately and what a great researcher he is. To me, it doesn't sound too fancy at all, but I don't really care. I am aware that my boredom is apparent, possibly oozing out of my every pore – I can't help it any more. About half an hour in, I can literally see his mood drop like a ton weight from a balcony. "How is your LpxC project going?"

He speaks sharply, pinning me down on the lab chair with his eyes. "Still struggling with the plasmids." *True statement but wrong answer, for sure.*

"We'll just publish what you have so far. Write me a draft for a paper and send it to me by Monday." *No way!*

"But that would go to a B class journal...," I stutter, seeing my one and half years' worth of failed experiments summarised in a magazine as glamorous as a primary school newsletter.

"We publish it, and you can focus on KdtA." *Oh yes, that megalomaniac project that doesn't have any future in our lab. Excellent idea.*

"Can't we just buy the plasmids?" I ask, desperately.

"No! Next week, I want it on my desk."

"But..."

"Get over it!" Mark barks, striding to the door.

Chapter 27

With pleading eyes Lucy says goodbye. "Don't do it," she whispers, standing in the doorway of my flat entrance; after a small dinner party to celebrate the acceptance of my LpxC paper in a low-impact scientific journal.

"Of course not," I say.

Lucy's parting look does not suggest trust. She knows, as she closes the door, Erico and I will be alone in the flat. She also knows that inmates of Lab 262 have been known to take refuge in each other's beds. Short affairs and one-night stands tend to be as far as it goes. Invariably, the net result is that the atmosphere in the lab actually manages to deteriorate.

Lab romances are commonplace throughout academia, of course. We are spending most of our waken lives in a small room together. Sometimes weeks pass during which our only serious contact with humanity is with each other. Frankly, we get bored. And we are not like pandas, which can exist in a cage for years sharing only shoots of bamboo. Or maybe it is an unconscious act of protest, akin to those promiscuous dissidents in former Bloc countries and Milan Kundera novels, whereby people affirm their freedom from brute authority and dependency by escaping into sex.

Less dramatically, it has a wearying inevitability about it. Babette and Barry had just proven that the scarcity of sanity in our bubble lowers expectations… Also their affair, which has trundled on for weeks now, has accentuated the atmosphere at work. Barry not only walks gloomily with his shoulders down now, but also as if he fears a shock stick from a jailer. He still talks to Mark and Babette, but to no one else. Sometimes,

© Springer International Publishing AG 2017
K. Bodewits, *You Must Be Very Intelligent*,
DOI 10.1007/978-3-319-59321-0_27

if Babette is not around, he exchanges a few sad words with Logan, but that is it. Babette is looking and acting more human than before, but she is still not talking to us – sex is only liberating to a point. Mark has by now figured out that they are dating and regards it as good reason to subject both of them to his monologues at once. Almost every day they sit in his office like Athos, Porthos and Aramis bolstering each other, by way of ground-breaking research plans, which will lead precisely nowhere.

"Really, don't worry," I assure Lucy as she closes the door.

I hear her walking down the stairs and walk back to the living room to face Erico. I am not attracted to him. He is a few inches shorter than me for a start. He's funny, no doubt, but simply not my type. He is changing the music from Amy MacDonald to blues. He turns away from the stereo and looks at me.

"Can I stay over?" he enquires carefully, though the words are somewhat pointed.

"You can, but nothing will happen between us."

"Can I sleep in your bed?"

"If you stay on your side."

"Why?"

"Because my sister told me that all Italian men have a small one?"

Erico looks confused and slightly irritated. "How does she know?"

"She had sex with one, and he had a small one."

I speak in a light clinical manner making it clear that I am not prepared to even engage on the subject.

"She slept with *one*?"

"Yes, one. But I don't see any reason to repeat the experiment."

"So I can stay over, but I can't touch you."

"That's the deal."

"Okay, I take it. I'm too lazy to cycle home."

I show him his side of the bed and change into a pair of shorts and a T-shirt. He undresses to his boxer shorts and gets under the blankets. I comment, "Sex between us would be perfunctory. I don't want it. I feel jaded and unromantic."

"Oh."

"Still, it's nicer to sleep together than alone, isn't it?"

"Well, normally when I share a bed with a girl more happens than just two people lying next to each other."

"It's the unusual things in life you will tell your grandchildren."

"You're weird, Ka."

"Who isn't when you get to know them?"

As agreed, he stays on his side of the bed.

With my coat already on, I stand in the doorway of my bedroom. "If you want to sleep longer that's okay, but I need to do a few things before my parents arrive."

Erico is taking a moment to remember where he is and how he got here.

"Ah, you told me your parents were coming, it's fine, I can nap at home."

"The one-night stand thing got a new meaning last night, didn't it?"

Erico smiles. "It did. I'm not sure if I will tell my friends about it."

It wasn't something to boast about, two young people not having sex in the same bed. It was friendly, downright familiar, but oh-so perfectly empty and low-key like the work life we were sheltering from.

I enter a small DIY shop on Atholl Place. An imposing, muscled guy wearing a wife-beater shirt and denims strides towards me. He looks around forty, though this may be an overestimate owing to his unhealthy demeanour and the splatter of old-fashioned tattoos on him. A particularly aggressive one runs from his neck up to his ear. A Harley Davidson belt and a leather body warmer would render him the quintessential Hell's Angel. "Can I help?" he asks in a coarse, Scottish accent.

"I'm looking for a mousetrap," I say very politely.

He leads me to the other end of the shop and points at all the mousetraps on a shelf. "Cool," I say and take one of the small pieces of wood with a little metal bar on top. They remind me of my childhood in the countryside; keeping mice out of the house during winter was a mini-war. Mr. Muscle regards me with unnerving suspicion.

"What do you need that for?" he growls.

"I've got mice in my flat," I say feeling oddly attacked; what else would I want a mouse trap for?

"Well, you don't have to kill them! Do you?"

He regards me with brazen contempt, like I am a defective member of society he is determined to march into the Maoist re-education camp he has recently reopened. I feel vulnerable to this condemnation. I don't necessarily want to kill the mice, but my parents are coming and I do not want to look like I have jettisoned all standards and happily live in squalor with them. I have a bad feeling that standards are precisely what this life is costing me. I want to at least pretend I am doing something about the mice.

"What's the alternative; isn't poisoning even worse?" I ask, pointing at the pink bottle with poison, recalling rotten, shrivelled bodies of mice in my parents' garage.

"You don't need to poison them either!" he says this with eyes wide open beneath distractingly thick eyebrows. *He really does think I am an awful person.*

"Here!" he says.

With his right arm, which is covered with a tattoo of a bleeding girl hanging upside down on a cross, he reaches for a small wooden box. He opens the lid and shows me what's inside; a labyrinth built from tiny wooden planks. It is beautifully made. It even looks quite fun for mice.

"You place some food here. They can only go in. They cannot get out. Every morning you check to see if there is one in there. If there is, you release it in the park."

"Does that really work?"

"You have to put it in the right place."

"How do I know the right place?"

"Put some flour on the floor before you go to bed. In the morning you will see where they walk. Put the wooden cage on their path."

"You're telling me I should cover my whole floor with flour?!"

"Yes! What you were planning was just animal torture!"

Feeling eminently bully-able these days, I reply, "Okay, I'll take that."

I pay six times more than I would have paid for a normal and probably more effective mousetrap. Mr. Muscle follows me outside and lights a cigarette in front of the shop. I open the wooden box before stuffing it in my backpack, feeling sorry for the mouse I will never catch, and cycle in the direction of George Street.

It is a cold day, and a winter glare makes me teary. I feel the mascara stinging lightly in my eyes. However, the sun is out and the air feels good in my lungs. From several spots on George Street you can see the Forth River just before it flows into the North Sea. There are numerous cargo boats making their way to and from the port. I wonder what goods they carry and who is crewing them. I stop to smoke a cigarette and have a few peaceful minutes overlooking the distant water. It's a spacious, dramatic view, and the tobacco smoke coming out of my mouth makes it even dreamier.

I feel a pounding in my head from the evening before. I was hoping I had avoided a hangover, but here it comes. My mobile rings in the pocket of my coat. *Who is phoning me before ten on a Saturday morning?! Who died?*

"Hi Ka, hope I didn't wake you?" says Logan.

"No, what's up?"

"The FPLCs are running dry. Seems you forgot to put the timer on and the buffer ran out. It has been pumping air for hours. I stopped them for you, but you might need to come in to fix it properly." *So that's it with the preparation for my parents.*

"Shit! I'll be there in twenty minutes."

I get back on my bike and cycle frantically through the Saturday crowds and traffic, up past Edinburgh Castle, over South Bridge towards campus. I am just locking my bike in front of the chemistry building when Felix walks out of the main door. He doesn't wear a coat and holds a pack of cigarettes in his hand.

"Hi!"

"Why are you here so early?" I ask.

"I couldn't sleep after 4:00 a.m., so I thought I might as well do something useful."

"Which is?"

"I'm helping a friend with application documents, and it's easier to do that in the office than in my apartment."

"Can't he write them himself?"

"He can, but it is a kind of psychological support as well. He's just in a grim situation."

I am pressed for time but I am curious about why Felix chose to fill in forms for someone else in the middle of the night. His friend's sorry story turns out to be a tale of our times... a scientist who has been post-doctoring for over six years; the last two years taking contracts in Switzerland which last no longer than two months each. He gets paid as a technician; far below his education level.

"Sounds like precarious seasonal work, except all year round," I say.

"Yes, it's fine if you're backpacking in Australia at eighteen but, being almost forty, having lots of degrees in your pocket and having devoted your life to science, it's scant reward. It's become clear he'll never make it in academia – he didn't get to publish enough."

"He only realises now that he's not good enough?"

"He has always been the best at everything he does. But he was unlucky in his last two projects. He wants to leave academia, but he's too old, too specialised."

"I get it."

Academia is a pyramid, and many are trying to get to the top. There is only room for a tiny fraction of PhDs on that little summit; a fraction, which is even getting smaller as the number of PhDs has increased rapidly in recent years. While one might reasonably expect to cling to the side, however far up you got, the reality is that as long as you are not secured by a permanent contract, you might roll off the side at any time – and it might be entirely beyond your control – and then you find yourself at the bottom of society. Having thumped to the ground, you are not valued higher than a fresh graduate, and sometimes less than people who have never gone to university or college. The shunned academic is seen as an age-defected grey-hound in a world that fetishes malleable youth. I had read and heard all that before, but so far it had been an obscure notion. Maybe we only sense the danger during our postdoc years, when it is already getting late… Or maybe this is stuff I do not like thinking about. None of us does. We try to ignore it and tell ourselves that we will be different.

Naively I enquire, "What about becoming a university teacher or a professor at a less famous university?"

"Even in Kazakhstan, the competition for professorships is fierce nowadays! And you know university teaching positions are like gold dust. They cut the money and the stability right out of the middle tiers."

All I can think to say is "Hmmmm."

"He'll find something but it won't be easy, probably a lowly job."

I listen sadly and Felix continues; he knows I am just waking up to these facts of life. "It's fine to work your ass off for a salary lower than a manual labourer on a building site if it leads somewhere, but it doesn't any more… We're all living on pretentions to the past. Now you can be deemed super-fluous at any point, and most of us will be… Sometimes scientists get stuck abroad when it happens, and aren't able to go back to their home country; people like me aren't part of the social system anymore, meaning no unemployment benefits back home if we fail, and only scant pensions, as we didn't 'contribute' for years."

Felix is obviously touched by the subject, as for a change he doesn't smile. He cares about his friend and is expressing fears that lurk in him as well. Frantic as I feel, I'm slowed down by the ominous weight of his friend's fate.

I pull myself together and rush off to the lab, where there is no sign of Logan or anyone else. I get into my lab coat, spin an extra scarf around my neck and go straight to the cold room where the FPLCs are standing. I can see straight away that air has got into every single tube. I prepare fresh buffer in the hope that the damage is reparable. My hands start to freeze from being at 4 degrees for too long but by then I am hopeful that the air is out of the system. My telephone rings.

"Good morning! You sound awake," my mum says.

"Yes, I'm in the lab."

"You do remember that we're coming today?"

"Yes, mum. I just needed to do something quick. Where are you?"

"I was calling to tell you we're at the airport. Your dad printed out the travelling schedule, and according to that we will be at your apartment in thirty minutes, okay?"

Aaaaaaaaaaargh! Thirty minutes! I need at least twenty minutes to cycle home and I'm not ready to leave! The flat is a bloody mess unless Erico cleaned it, in which case I would invite him to stay over every weekend.

"Sure."

"I hope you're ready for us."

My head starts spinning and I consider trying to delay their arrival – I can think of no tricks, options or lies. I bolt downstairs to the Simpson group and ask Felix to exchange the buffer for me in half an hour.

"No problem, off you go!"

I rush home and calculate that I have between five and seven minutes before my parents ring the bell. I collect all the empty bottles of beer and wine from the evening before and chuck the lasagna stained plates in the sink. I wipe the kitchen table and benches at vigorous speed as if it were a competitive sport. I throw all material stuff in a wash basket. I take out the hoover – a rare event, hence the mice no doubt – and race it over the floor. Just as I almost finish hoovering the living room, the doorbell rings.

"Hi, I'll buzz you up!" I say, trying to sound leisurely.

The time it takes my parents to walk up the two flights of stairs I spent spreading flour over my wooden floor and placing the mouse labyrinth in the middle. It's a clever touch, looks organised and thoughtful – the handiwork of someone with standards.

"Hello," my mum says overly excited, before giving me three kisses on the cheeks.

She looks far too fresh for having left their house at ridiculous o'clock this morning. My father stands in the doorway behind her, holding something that looks like a large bag containing an electronic device I probably don't want. I don't like being burdened with large possessions.

"Can I go through, so I can set this down?"

"What is it?" I ask, trying to sound cheerful.

"A computer. It's not new, but it works. And the university can install whatever they want on it to get you online."

"Oh wow. That's great!" I say, genuinely excited.

Now I have a computer as well as a desk. So reduced am I, these two commonplace items stir feelings of enormous well-being deep inside me.

"Thank you so much!"

"Hopefully it works, because they weren't happy with me bringing it in the plane."

„ ...these two commonplace items stir feelings
of enormous well-being deep inside me... "

"Coffee?"

"That would be nice," mum answers.

I jump over the flour to get to the sink.

"What's that for?" she asks, pointing at the flour and the cage.

"It's a mousetrap."

"What's the flour for?"

"That is just to check if I placed it in the right spot of the room. The mice leave footprints in the flour."

"Does it work?"

"Not so far."

"Shall we get you a proper trap?"

"No. I want to try this first."

Parents are a lifelong touchtone with your own reality. Right now, my reality consists of trying to impress my parents with how grown up I am. But only this morning I was bullied into buying the mousetrap I didn't want. My experiments are going awry and, even worse, I'm not sure it matters since my job – my supposed *raison d'etre* – is probably going nowhere in reverse gear like Felix's friend in Switzerland. I share beds with colleagues I do not want to sleep with. Of course, I don't reveal any of this. I successfully impress my parents the way one is supposed to, while seeing myself as just another desperate, phony pseudo-adult in the crowded rut of academia.

Chapter 28

"Ka, I need to talk to you!"

Vlad is on the phone sounding even stranger than usual; worried, or excited, I really can't tell.

"You okay?"

"I guess so."

"Is Daniel okay?"

Thankfully, I haven't heard from Daniel for months, but as far as I know, Vlad is still in touch with him.

"I don't know. I haven't heard from him for a while. I saw on Facebook, apparently he is a bike courier now. Perhaps he was hit by a garbage truck and is dead."

"Yes Vlad, very Russian. Where are you?"

"The airport."

"Where are you going?"

"Nowhere. I just came back."

"Ah yes, a research trip…"

Vlad is always on alleged research trips, all over the world. He is extraordinarily adept at misappropriating funds. Brazil, Namibia, Iceland, USA… you name it and Vlad will have convinced some funding body in the scientific world that he needs to collect some random samples from a deserted beach in that country in order to continue his ground-breaking research.

He must be a wondrously talented grant writer as his publication output from the University of Edinburgh is precisely zero. Given that his supervisor has not published for the past twelve years, it seems fair to assume that his output will remain non-existent throughout his PhD. It is a shame because

© Springer International Publishing AG 2017
K. Bodewits, *You Must Be* Very *Intelligent*,
DOI 10.1007/978-3-319-59321-0_28

Vlad – despite some socially challenging behaviours – is a remarkably intelligent man who could perform meaningfully in another environment. Such is modern academia that his intelligence is squandered on the debatable art of grant sponging. Mostly he travels alone but occasionally he takes along a random girl he met online.

"Are you in trouble?"

"No! At least I don't think so, I'm not sure… I can't tell you by telephone."

He laughs, which would be reassuring, were it not for the strange pitch of his voice.

"Are you at home?" he asks.

"I am."

"I catch the next airport bus and will be there in half an hour or so."

"Shall I come to Haymarket to help you carry your bags?"

"Women don't carry bags, Ka."

Less than an hour later, Vlad drops off his bags in the little hallway of my flat and sits on one of the red benches in the kitchen. He looks at the silver strainer lying upside down on the wooden floor.

"Do I see that right? There is a mouse in there?"

"Yes! I caught it! Just before…"

"You caught it?" He sounds astonished.

"It took me a long time."

"And what are you going to do with it? Kill it?"

"I don't know. I am not going to kill it. I thought…"

"Throw it out of the window!" Vlad says slightly disgusted at the thought that he's sharing a room with a mouse.

"Vlad… I live on the second floor. I am not throwing mice out of the window. Apart from the fact that it would be animal torture, what would the pedestrians think about flying mice? "

"It is only one, Ka."

"You think one is fine?"

"Yes." Vlad looks entertained.

"I would rather take it to the graveyard on the other side of the road."

"Oh, come on! Would it not be funny that you walk on the street eating some disgusting fish and chips and all of a sudden there is a mouse falling into your food?"

"Hilarious. Why are you *here?*"

He doesn't answer and looks like he isn't sure anymore whether to tell me or not. I fill up the kettle to make tea and repeat: "Why are you here? Do I need alcohol to handle the story?"

"You probably do."

I open the fridge and take out an opened bottle of fruity white wine that has been sitting in there for a week. "It is really bad wine, so your story better be good."

"Is a Russian causing an explosion in a hotel in Japan during the G8 Summit a good story?"

"You caused an explosion? You? An explosion?…"

"Yes."

I sit opposite, leaning my elbows on the kitchen table, staring at him with wide open eyes – anyone else and I would think this is nonsense or exaggeration. But not with Vlad. He looks back at me, blankly.

"Tell me!"

"Well, you know that I often go on trips."

"I am aware you suck all funding pots dry."

He sighs: "If they are stupid enough to grant it to me…"

"Just tell me about the explosion."

"So, yes, I get all those samples from beaches and sometimes I need to take samples to places as well. I transport them in dry ice, you know, in one of those huge Styrofoam containers. It takes a lot of space in my bag and that started to annoy me; I could never bring nice souvenirs, for Tatjana and Lucia."

"Recent targets from the Internet?"

"Yes. So, during my last trips, I decided to bring a travel mug instead and hope for the best, that it would all arrive still frozen. It worked, but not super good."

"Like… they arrived at room temperature?"

"Yeah, something like that. So this time, I decided to buy a high quality travel mug – a very nice Japanese one. I filled it to the top with dry ice and my samples. I travelled with my coffee mug from the lab I visited in Osaka, which took a few hours. Late in the evening I booked into a hotel, placed the mug in the minibar fridge to keep it as cold as possible and went to bed. The next thing I can remember is a big bang, a fridge door flying over my head into the wall and other random pieces of god knows what flying around. It was so loud and such a mess! I was scared… really scared! I had no clue what was going on until I saw the coffee mug stuck in the plaster in the corner of the room. I tried to pull it out, but it was impossible. I tried to find the samples, but that was also hopeless in the dark. Then the next thing that came to my mind was I am a Russian building a bomb in Japan during the G8 summit in a hotel where I checked in under the false name. That will be a few nights of prison!"

„ The full dry ice experience. "

"You checked in under a false name?"

"I know; sounds strange."

"Not at all, why would anyone use their real name – so boring!"

I take a large sip of the horrible wine.

"Yeah, that was pure coincidence. I paid the hotel in cash at arrival, so they did not need to see my passport. I just wrote something… Anyhow, I decided to run because the whole hotel would be looking for what was going on. I took my bag and hurried nine floors down. Nonchalantly I walked out of the hotel and continued walking until I managed to stop a taxi. He drove me to the airport, where I waited seven hours for my flight to take off. I was so afraid the Japanese police would come to fetch me. Paranoid!"

There is a moment of silence in the room. Vlad sips from his tea, I drink my wine.

"The explosion could have happened on the plane!" I state.

"Yup. Scary, isn't it?"

"Yes."

"You won't tell anyone, will you?"

"I didn't sign a confidentiality agreement, did I?"

"Please."

"No worries. Take the mouse to the graveyard and I won't tell anyone."

"I am not bringing that mouse to the graveyard! All I can offer is to throw it out of the window."

"No! Leave it. I would love to tell people though. It's a good story."

"You can tell my sweetheart, Lucy. Maybe she marries me thereafter."

"Oh yes, definitely. A done deal, I'd say…"

He looks sad in exactly the same childish way as Mr. Bean looks sad. "What will you tell your boss?"

"Nothing."

"What if he asks about the samples?"

"He won't. He is too busy teaching Benjamin, the chubby Chinese noodle chap, to write his name. And the research doesn't matter to him – of course."

Chapter 29

Frantically, Barry looks at the list of telephone numbers hanging on the wall next to the office door. It hasn't been updated properly for years. I scribbled my number at the bottom and Logan wrote vertically along the side, and so on. Barry isn't in the lab much lately, but when he is here, a student is invariably hanging on his tail. He is burdened, by Mark, with one student after another, and sometimes several at once. Today he is alone, looking for one of his student's phone number. "You've got an hour free?" I joke.

Before his reply, Barry quickly scans to see if his lord and master is around, but Babette is absent.

"Yes, finally. I can't wait till term finishes and I can focus on my own work for a bit," he says, clearly uncomfortable about talking to me; possibly an act of treason under Babette's rule of terror.

"I wouldn't look forward to it too much. Mark told me this morning that there will be a German student coming for a summer internship. He didn't say who'd supervise him, but it might be you."

Barry shakes his head and slaps the table. "Fuck!" *Wow, that looks very unnatural, Barry just ain't dude-ish enough to let loose with attitude…*

He recovers from his gesture, looking like he might have scared himself a bit with it. "You've got Gamon's phone number, per chance?" he asks.

His teeth chatter. His panic is sincere. Oily-looking sweat is visible on his cheek bones. He knows the answer to his question. He is clutching at desperate hope. And all for a student's phone number?

"No, sorry. What do you need it for?"

Barry stares out of the window in the direction of the empty parking lot. It is a lonely view with just a few trees blossoming at this time of the year.

© Springer International Publishing AG 2017
K. Bodewits, *You Must Be Very Intelligent*,
DOI 10.1007/978-3-319-59321-0_29

Barry moves his eyes from the parking lot to my face, biting his tongue before he replies: "He is giving a presentation about his work at midday."

"Oh. What work?"

"That's what I wonder as well," says Barry, somewhat tightly. "And that's why I would like to see his presentation beforehand."

I refrain from asking any further questions, for fear of falling into the same argument we had before about Gamon. He is an overseas student who, for six months now, is supposed to be working full-time on a research project in our lab, in order to complete his master's degree. He introduced himself to Mark and visited the lab the week before he was supposed to start. He didn't show up on his first day, or his second or third… After three weeks, Mark announced that Gamon had sent a message saying he would finally be starting, the next day. The reason he had not shown up thus far was that unfortunately he had been unable to find our lab. Mark laughed it away, in a rather forced style. He did not announce any sanctions for this blatant lie coupled with incompetence.

I had made an off-hand joke about it, "He couldn't find the lab he had been to the week before? He is either demented or as vital and motivated as the decaying corpse of a long-dead sloth…"

"The decaying corpse of a long-dead sloth?" Logan had repeated my words in disbelief, resting his head on his hand.

"Just saying: Whatever it is, he doesn't seem suitable for a master's."

"He is a paying customer," said Mark, dry and business-like.

"He's not a customer, he is a student!" I blurted.

It bores me having to treat undergraduate students like precious princes and princesses. Some seem to be allowed to do whatever they want, including showing up for practical classes only when the mood takes them. I have often had to stay in the teaching lab after hours – sometimes _for_ hours – just because students "rock up" late. Really weirdly, their exams are not their responsibility; it is, apparently, our problem if they don't pass. We have to make them pass no matter if they are good and motivated, or completely hopeless oinks. Alas, some of them are good at one thing – filing complaints, or at least threatening to do so, at which time we run after the deceitful little brats like fawning office juniors.

For some it seems that if they enrol at the University of Edinburgh, they have completed the only vital part of the course already – job done, time to party. The degree will be yours unless you refrain from "rocking up" at the graduation ceremony. Actually, even then you will get your certificate by post. I do not think this is unique to Edinburgh; I fear it is the way much of paid-for academia is going – down the plughole with my ex-student's vomit. It is sad. And it is down to money-grubbing of course. The students who get away without studying are overseas students paying for their course. Kicking them out after a year or two – and there are some who should cheerfully be kicked out after a week or two – means the university would forego another year or two of hefty tuition fees from their rich foreign parents.

"Gamon is coming to our lab, Karin, starting tomorrow," Mark had insisted.

"It will be a disaster."

Mark leaned forward on the table, into confrontational mode. "Students from abroad pay very high tuition fees, they keep the university running. We need them and they want to go home with a degree…"

In all fairness to Mark, his tone clearly stated; "Get off my back. I know it's a pile of immoral, greedy crap whereby lazy, thick, rich brats get awarded degrees they don't deserve in a month of Sundays. But I can't do anything about it. And neither can you!"

„ He is a paying customer, he will get his degree. "

Barry didn't object when he was told he would be supervising Gamon. Barry never objects to anything. Now six months have passed, or rather five months and one week because Gamon never had to make up for the three weeks he spent searching for the lab. Since that three-week struggle to find a room any fool could find, Gamon has graced us with his presence roughly once a month. At one point, I suspected he came into the lab at night. When I would leave last, late in the evening, and then be there first in the next morning, I occasionally got the feeling someone had been here, in between times. An ice bucket standing somewhere I could not remember leaving it, a light left on, or a window open... I could never truly say for sure that someone had been in, because the mess in the lab is hard to keep tabs on. But I'm not the only one who had the impression that our lab was being used in the small hours. Lucy and Logan had also remarked on it. And we all wondered if it could be Gamon fiddling around. In his country the clock runs nine hours ahead so maybe he never adapted to the "local" time, we speculated. It is a spooky thought, sending faint chills down my spine; not only because there is someone working alone in a chemistry department in the deepest night, but also because he is a student who does not have a clue what he's doing with tricky chemicals.

Barry leaves the office, looking desperate. "Good luck!" I say.

As soon as I hear the lab door fall close I ring Lucy, "Where are you?"

"Glasgow. I told you yesterday. Mark insisted. It's to work on 3D pictures of my proteins."

A short silence; I'm disappointed she is not here.

"Why?"

"No one is here today, no people in the office. It's a bit boring."

Almost nine months after their stipends ran out, Hanna, Quinn and Erico had left Lab 262. If there wasn't a university rule stipulating that students must submit their thesis within four years of their starting date, I guess Mark would not have let them go. I recently learned that – thankfully – the School of Chemistry is hitting its staff hard for crossing the four year line. What "hitting hard" exactly means is unclear to me, but in my imagination naughty supervisors get their bare buttocks publically whipped in the large lecture theatre (I know that's a long shot). But whatever it is they do, it caps the phase of unpaid slavery at twelve months; a protective measure many PhDs around the world don't have.

Without Hanna, Quinn and Erico, it is quiet in the lab. There is just Logan, Lucy, myself and the project students. At the start of this academic year, Mark did not take any new PhD students up. I know not why, but assume either money or moodiness got in the way. Had Linn, an undergraduate, not announced she would like to stay on for a PhD, we would not had any new blood in the upcoming year either.

The lab is dead today. I love seeing Lucy any time but today I arrange evening drinks with her simply for the sake of human contact with anyone.

The timer beeps on the desk in front of me. It is time for the next experimental step. I head to the lab and sit on an empty bench. There is a tray in front of me with Eppendorfs numbered 0-23. I open the lids and pipet a few microlitres of Coomasie Blue solution into each of them. Four hours and a lonely lunch later, Barry comes in. "Did Gamon show up for his presentation?" I ask.

"He did."

Barry sounds relieved.

"How was it?"

"Strangely enough it was excellent. He got very good results. Very clean gels, he purified a difficult protein, and he presented it well."

I can feel my eyes widening while he is talking.

"That can't be!"

"Well, apparently so." *You don't actually trust this, right?*

"That can't be!" I repeat.

"He must have done it during the night."

"Seriously, Barry, IT CAN'T BE! An undergraduate cannot purify a protein during the night without help. And no one ever showed him what to do. I just don't believe it."

Barry shrugs his shoulders.

"It doesn't matter, he's finished now. Back to Thailand."

"With his undeserved master's degree…"

Barry walks to the fridge, digs through a few boxes and takes some samples out. He holds them against the light.

"This should contain the protein Gamon purified," he says.

He takes a microlitre out of one tube to quantify it in our newest piece of equipment – bought from Barry's grant – the nanodrop. He presses a few buttons and looks at the display. His smile is bitter.

"It's water, isn't it?"

"It is."

The late evening sun is projecting shadows of fluttering leaves on the wall of the lab, in exactly the same patterns as it did last year of course, but it's no less gorgeous for that. Lucy finally arrives, two hours later than she expected.

We walk over to KB House where lots of undergrads, PhDs, lecturers and professors are celebrating graduations and the end of the academic year. For me, another PhD year has passed and I am determined that no matter what happens, I will stay no more than one more year in the lab. Then the practical part is over and done, and hallelujah, I will write my thesis.

This is an ambitious plan given that the research results garnered thus far are, to put it as politely as possible, a bit crap. The entire last year has yielded nothing beyond a very short paper based on the results of my first year. Reading that paper, a reader might reasonably think it had just been a little side project, with sweet results of a smoothly ran antibiotic assay on superbugs; worth publishing purely as an end in itself, a minor point of interest for the super-diligent student. That would be to disregard all the hours, days, weeks and months I have spent desperately trying to prove a hypothesis about why some bacterial strains are susceptible to the antibiotic and some are not. It doesn't reveal how I have been drifting from one disastrous experiment to the next or how often I have banged my head on the table out of desperation; or how many bottles of wine I downed on the window sill with Lucy or alone to just escape the reality. Scientific articles show success, not failure. In a way that's all good with me – that's science, that's endeavour and that's why the breakthroughs are so rare and beautiful. However, I also know now that in this lab, with this supervisor, I don't have a chance of success.

The thought of leaving the University of Edinburgh to re-start my PhD elsewhere crosses my mind regularly, but I know it is fantasy. Leaving and failing is not an option. I have lived into hope too long. The investment has been too big – time, money, time, relationships, time, and wine – I want a doctor title in return. Plus, I like Scotland. I like Edinburgh.

My relationship with Mark has neither significantly improved nor worsened – it's been flat-lining along in desultory style. He stopped helicoptering me and adopted an extreme form of "free-range supervising," meaning that most days I don't exist for him, or at least am no more significant than a budgerigar's daydream.

It's been fine. More and more, Mark has been sucked into an obsession with the KBL project, existing in a troika with Babette and Barry, and taking little or no interest in the rest of his postgrads. In the past, to vent steam, stress and frustration, he shouted mostly at the senior PhD students; Hanna, Quinn or, very occasionally, Erico. But now that they are gone and I am the next inmate in line, I know it is just a matter of time before I get placed in the pillory.

I order myself and Lucy a large glass of wine and an alcohol-based warmth soon infuses my biological body. Methinks I've been conducting this experiment too much of late but, hey-ho, it stops me fretting about being pilloried throughout the coming year.

Part IV: Year 3

Chapter 30

"Dead," Lucy says in a shocked voice while staring at her old-school computer screen.

With my coat and backpack on, I pause at the office door. *Who is dead?*

"What's going on?" I hear the tremble in my voice.

"I just told you, it's dead."

She keeps her eyes on the computer screen. "*It* is dead?" I ask, taking a seat beside her.

"The cat."

"Which cat?"

"Whatsoever."

"Whatsoever? You mean the virtual cat?"

"Yes. It's dead."

"Fuck Lucy. You scared me!"

"I scared you?"

"Yeah Lucy. You sit here... with the body posture of a bag of potatoes, staring at a computer screen as if it's nine-eleven and you greet me with the words IT'S DEAD."

I should have stopped there, of course, but I riff on. "I am just off my bike, I actually enjoyed the morning sunshine! I was expecting a normal boring lab day. You know... like... you sit on that bench and assay proteins and I repeat the same experiment for the tenth time without success.

© Springer International Publishing AG 2017
K. Bodewits, *You Must Be Very Intelligent*,
DOI 10.1007/978-3-319-59321-0_30

Babette sits there emitting strange, grumpy noises and Mark storms in at ten to vent his daily darkness. We would try to hide from him for a coffee break. At twelve we would go for lunch at KB House where we eat a hamburger with potato wedges... and then we would go for a cigarette break every hour in the afternoon, and wonder if all life is pointless. I expected it to be normal. Of course you scared me!"

Lucy grins. "Stop Ka."

But I don't stop, why would I?... "I thought you would be flying to Belgium, to bury someone!" I add. "But what happened is that your virtual cat died?"

We are both laughing now, though I still keep a theatrical expression on my face as if she had really screwed me over.

"Sorry," she says. "I didn't know you would be so sensitive about it."

It was just over two months since the end-of-year party at KB House, when Lucy was offered a virtual cat by one of the lecturers in the department. I saw him talking to her for a long time, and I couldn't imagine what they would be talking about so intently. It was nothing new, Lucy being captured by a random male, but this one didn't let go of her for ages.

"OMG," Lucy had said when the email came in from the virtual cat app.

"What's that?" I had asked.

"You remember this guy... from last week, Dr. Wilson from the inorganic chemistry department? He offered me a virtual cat!"

I felt my lips struggling to avoid a smirk. This was a new one. I loved all those guys proposing and doing strange things to score Lucy. I love all those messages, songs and poems. I get her to read out every single one, sometimes over and over again. I feel like a little girl sitting on the sofa replaying the end scene of *Dirty Dancing* ten times in a row, and wondering if this time Patrick Swayze will finally lose control and catapult the chick against the wall. The way Lucy reacts fascinates me. Despite being pursued by half the planet, she still reacts with honest surprise, and it's generally shackled to a quite profound disinterest in whatever guy is trying to score her. I guess any other girl would become arrogant but Lucy never does. Every single time, she is surprised and yet sanguine.

"How come he's offering you a virtual cat?" I had asked while stifling a snort.

"He asked me what my favourite animal was. I said pigs and started on about the variety of sausages I like. He clarified very patiently, that he was meaning pets..."

It didn't take long before I had felt the tears running down my cheeks. Guys around Lucy – they just lose the plot entirely.

"Did he ask you about your favourite colour as well?"

"I can't remember, I was really drunk that evening. But I remember telling him that I like cats and that I would love to have one, but that it is just not viable right now. He told me that he felt the same, and would also love to have a cat one day."

"What a connection!"

"It was the *only* connection."

"Are you going to accept it?"

"Of course," Lucy replied, already opening the link of the app he had sent her.

"I still need to give it a name."

"Give it a name which clearly shows Dr. Wilson you are not very interested," I had advised.

"Whatsoever?" she had proposed.

As soon as her black and white cat appeared on the screen, we were informed that Dr. Wilson himself keeps three virtual cats and a dog. The instructions said Lucy would need to feed her own cat regularly and give it attention. By way of touch screen or buttons, you could give the cat food, water or milk, and you could cuddle or pet it. You could do the same with other people's virtual cats. Dr. Wilson must have gotten an email saying that Whatsoever has been registered as he cuddled and fed it within a few seconds of Lucy signing up.

"He's doing stuff with it!" Lucy had said in disbelief.

"Oh dear… he plays with it!?"

Over the summer, a few times per day, there were emails coming in saying that Dr. Wilson had cuddled, petted or fed the cat. Of course we wondered what Dr. Wilson was doing all day that he had so much time left to care about his own virtual animals and the cat of someone else. After a week or so Lucy had started to get annoyed, and after two weeks she hated it, she hated the messages in her inbox, and actively looked for buttons to get rid of the cat.

"I can't hit the cat. I just want to hit it."

"Hit his cats instead," I suggested.

"There is no button for it either." She sounded seriously aggrieved.

"Can't you starve it?"

"He keeps feeding it. Why do I only attract idiots?" her eyes rolled in exasperation.

"This Jörg was alright, or…?"

„ Oh dear, he is doing stuff with it."

"You mean the dude who – on the first date – served me days-old pancakes in chicken broth and called it pancake soup? Only Germans do such disgusting things."

"Yeah, not a promising start."

Every day, Lucy kept me updated on what Dr. Wilson was doing to her virtual cat and we mostly laughed about it despite her annoyance. But in recent weeks we had lost interest in the cat, and so had Dr. Wilson, apparently. He has been feeding and cuddling less and less, scaling down to just five times per day then three then one and finally stopping altogether. And now the time had finally come: the cat was dead.

"You should go to his office, to cry because it is dead," I say.

"Yes, I kind of presumed that he would feed it."

"Right, you're not seriously going to Wilson's office to cry," Logan asks, baffled.

"Well, it is kind of his fault it's dead, isn't it?" Lucy defends.

"He could have put more effort into keeping it alive," I add.

"Yeah, he could have."

Chapter 31

"Ohmigod, ohmigod, of course we go!" I shriek, in the office for all to hear.

"And I thought Jennifer Aniston was hysterical," says Lucy drily.

"I would just *really* like to go!"

"You *want* to go the Oxford University alumni party?" Lucy asks, in a distinctly disparaging tone.

"Yes!"

I am picturing myself sipping cognac with Alex amid the über smart set within a filmic university setting...

"You fancy Alex, don't you?"

Lucy's beautiful green eyes do their eerie penetrating thing. I think about his charm; his brown eyes, his defined mouth and despite his body not leaving much trace of physical activity, the appeal it has. There was something about him, as if he were designed to revive with my stone cold dead fantasies about Romantic academia in the UK. "I guess I do," I answer honestly.

Lucy's beautiful eyes roll around as if despairing of this unfathomably strange statement. "He is charming," I add.

"Handsome he is, but mathematicians crunch numbers, not mattresses."

Did she just say that? Lucy of the devout demeanour is never quite so crude in public...

"Did you just say that mathematicians crunch numbers but not mattresses?" Logan asks.

"That's what I understood," I say. "Possibly a good point."

© Springer International Publishing AG 2017
K. Bodewits, *You Must Be* Very *Intelligent*,
DOI 10.1007/978-3-319-59321-0_31

"You girls are awful," says Logan, shaking his head and walking away to turn the autoclave on lest he hear more.

Alex and Chris, both professors-to-be in the School of Mathematics, had been at my flat party the previous weekend. I had invited them during our first encounter at KB House a week before. We had been having after-work drinks that evening, which by now hardly merits mention since that is a normal evening. When eleven rolled round, Lucy and I were both clutching our empty glasses, holding onto hope while we treated our new maths pals like long lost friends. They dutifully volunteered to buy us a beer and we graciously accepted. It's the acceptable indignity of the PhD experience; being skint when you've reached an age where you shouldn't be skint. It has to be endured, lest only the rich reach the upper floors of the Ivory Tower.

Earlier I had spotted a gorgeous guy at their table – Alex. When Lucy and I went outside to smoke a cigarette we were joined by a tall, long-haired PhD student, Greg, also seated at that table. "Who's the chap in the corner?" I had asked.

"My supervisor."

"Oh."

"He didn't offend you, did he? He doesn't seem to get along with the French, or philosophers. Or many people really. So don't take it personally."

"No, he didn't. He's sitting too far away, haven't exchanged a word."

"Keep your distance." _Oh, mysterious…_

"You should be glad you have such an extremely good-looking boss."

"I don't know in what way I benefit from that."

"Isn't it nice to have someone aesthetically pleasing around you? I love being around Lucy, I love her as a person but I also love looking at her; the shock of her beauty recharges with every glance."

"Ka… you are drunk!"

"I don't think heterosexual men savour each other's beauty the same way," Greg had said.

"He is sooo hot!"

"His PhD student is standing here," Lucy reminded me.

"You are all about looks?" Greg had asked.

"No, not at all," I protested.

Lucy and I gave Greg the once-over; not a guy who put any effort what-soever into looking good, or even dignified. The T-shirt looked like a freebie from a local sports event and the jeans could have come from a pound shop.

His cycling shoes were cool, if only he were in the middle of the Tour de France; for a city commute they were simply silly.

"Are you going home?" I had asked him.

"No, what makes you think that?"

"You have your jeans rolled up for cycling."

"True, that's so they don't get caught in my bike chain."

"Yes. I guessed that. But you're not cycling now. You're having a beer, right?"

"Yes, I'm having a beer right now. Would it please your sensitive eyes if I were to roll these down ma'am?"

"I guess it would," I said.

"It definitely would," said Lucy.

When KB House closed its doors around midnight, I invited all three left-over mathematicians to my flat party. The next morning in the lab Lucy asked if I recalled this dubious display of largesse.

"I remember, yes, but they won't come."

"They will. Being invited to a party is a rare event for them, maybe the highlight of their year."

"They don't even have the address."

"You gave it."

"Oh. Well, they won't know it anymore."

"Believe me, those dudes process data like computers."

Lucy clearly has a downer on mathematicians, but she was right about their attention to detail, because there they were; Greg had shown up in his lemon-yellow bike jacket, with his PhD supervisor Alex, and Alex's buddy, Chris. When I opened the door, I said "Oh," perhaps a little too excitedly.

There are three weirdos entering your party, be friendly and don't make a fool of them or yourself. But by God he is still gorgeous...

"The mathematicians have arrived; the party can get started!" I shouted through the flat.

Greg and Chris hadn't been offended but Alex seemed to be considering my clichéd joke very carefully, maintaining a calm expression of complete self-assurance. *Weirdo (But a gorgeous and mysterious weirdo)...*

I gestured them in but Lucy was at the door too, pointing at Greg's rolled-up jeans. Without saying a word, she was telling him to roll them down before entering the house. Her beauty is commanding. There is never a doubt men will obey. I, on the other hand, have to resort to words, "And maybe take your shoes off as well."

„ The mathematicians have arrived,
so the party can get started! "

He did as I asked and on display now were his large feet in pink Pilates socks separating all five toes. Lucy stared at the socks and then the jeans – same shapeless cheapos as last week. Her eyes moved up to the T-shirt; another one that can only be a freebie – it is inconceivable that anybody would pay actual money for such an ill-fitting, flimsy piece of sponsorship cloth.

I try to apologise, "I'm sorry. I was concerned you might damage my wooden floor with the click system on your soles. I did not intend to expose you as a tasteless lunatic."

I catch Lucy's eye, and she presses her hand to her mouth to keep the beer inside, then rushes to the bathroom to spit out and burst into laughter. She doesn't want to hurt the man's feelings but he dresses as if he actually wants to be mocked.

"A girl who makes it clear what she thinks," says the mysterious Alex, hanging his jacket carefully on the overflowing coat stand and, finally, he smiles.

"They're very comfy," Greg defended his indefensible socks, then laughed. "You want me to take them off?"

"Nooo! Keep them on. They're a party piece."

Alex and Chris take a dozen frozen bottles of Schnapps out of their backpacks and try to fit them all in the small freezer. "You guys have big plans," I comment.

Alex's complete non-response suggests either he is deaf or I am deathly dull. *Is it normal to bring so much booze?*

There are about twenty-five people now, invariably PhD students from the biology or chemistry departments. Alex and Chris are obviously the oldest guests, and they draw some sideways glances.

Logan stands in the kitchen doorway pointing at the backs of Chris and Alex. "What?" his lips say without making any noise, looking at me with a big question mark in his eyes.

I walk towards him. "What are they doing here?" he whispers.

"I invited them. Do you know them?"

"Of course! I was in their lectures as an undergrad."

"Cool."

"It's awkward, Ka. How do you know them?"

"Lucy and I met them in KB House."

"Right…"

"So who are you, Karin."

Quite suddenly Alex is right behind me.

"I'm doing my PhD in the chemistry department," I say, slightly nervously I fear.

Why am I so desperate to impress this one strange person? And when the opportunity knocks I know not how to answer…

"I'm aware of that. You work for Mark McLean."

"You know him?"

"I do. How do you like it?"

My mind is flying. Does he want to know about my PhD or about living in Edinburgh? How well does he actually know Mark? Should I blurt out that my PhD is a nightmare and Mark is a jerk-off? Should I say my PhD is as intellectually stimulating as Kim Kardashian on ketamine? Perhaps I could tell him how horrible the afternoon was, with Mark shouting at me, again, it's virtually a daily ritual of late. Today was typical; me not having my priorities right, focusing on the wrong project blah blah. But the number of projects and the importance of each were redefined, as they are every few days, and the only consistent fact about them is that none are remotely realistic because of time and/or money; I only have one year left and our threadbare lab simply, in reality, does not have the necessary equipment. The shouting shreds my nerves till I hardly have a clue anymore what I'm doing or should be doing. But I can't tell Alex about my personal situation, how the hell has the idea even occurred to me, how desperate do I feel?! I don't know him, maybe he and Mark are best friends, though I'm not sure Mark does friends. Maybe he plays football with Mark, though somehow Alex doesn't seem the footballing type, too stylish for the compulsory coarse banter... Maybe it's a trick question and he's trying to make me say something bad about Mark, which he will then convey to him. Maybe... I thought all this through in nanoseconds, and then I came up with the brilliant conversational gambit, "It's okay."

His eyes are quizzical, perhaps trying to work out how someone can seem so alert and yet speak so dully. Brightly I add: "I like Edinburgh very much."

Oh aren't I witty fun?!...

"I'm glad you like it," says he.

I walk off to the fridge to get myself another bottle of beer, perhaps hoping to find some words and confidence in it.

Lucy follows me. "It is possible to have a conversation with him?"

"Not sure. He makes me nervous."

"Greg warned you that he can be quite offensive," Lucy said, worried.

"He didn't offend me. There is something about him."

I walk to the living room. Both Greg and Chris have integrated into the party, happily chatting with all and sundry. Alex sat at the kitchen table, keeping company with whoever chances to sit down.

Just after Vlad left, Alex got up from the kitchen bench, apparently about to leave too. (I later learned this was after offending everyone who came into his orbit.) It must have been around 3:00 a.m. when I opened the front door for him. He pushed me away from the door and pulled me to the back of the corridor, and held me before my bedroom.

"So you think I am hot?" he asked in an accusing tone – sounding more like he was asking if I really called his mum a hooker. *Stupid Greg.*

I stood there, flabbergasted and unable to take my eyes off him. Little shards of grey were creeping into his brown hair, his shirt was perfectly ironed, his jeans were so-so but he could carry them off with ease... so extraordinarily handsome, so shockingly self-confident. He was examining me, gauging my reaction. I guess if I had been sober the situation might have scared me. But I wasn't sober and took a deep breath. "Yes, I do think you are hot."

"So then why don't you kiss me?"

"Probably around twenty per cent of the worlds' population would agree that you are hot. It might get a bit disturbing if over a billion people started kissing you on the basis that they found you hot."

"How do you come to this twenty per cent?"

I was trying to sound unfazed but – Jesus Jones! – NEVER give a random number to a mathematician...

"Eh?... It's just a rough estimate."

He looks disappointed by my absence of algorithmic calculations based upon sophisticated models of mathematical probability.

"It includes two per cent males," I add.

He looks slightly disturbed. "Just bloody kiss me," he says.

So I did.

As soon as I stepped forward he gripped my lower body and pressed himself against me. After only a minute or so he told me he had to go. He had to catch a plane, flying to Brussels. I didn't ask what he would be doing there over the weekend, no need; Greg had told me earlier in the evening that he had a girlfriend in Brussels. I didn't care. I had snogged the probably hottest almost-Prof in the School of Mathematics. Maybe even the hottest academic at the University of Edinburgh?... With a few drinks inside, that feels like an achievement, much more of an achievement than acquiring the desultory PhD that consumes my days.

My life really has come to this; in my mid-twenties I measure achievement in drunken snogs because, sober, I know I am destined to achieve nothing of any significance whatsoever in the science lab which drew me to Edinburgh. That is just a sentence being served. And inmates often turn to narcotic escape and ill-advised sex...

I walked back to the living room, joining the remaining guests. Towards the end of the party Greg was still there, fully integrated with the remaining few chemists and biologists; a surprisingly outgoing and socially skilful guy despite the buffoon socks.

"I snogged your PhD supervisor" I said, trying to sound nonchalant.

"You... you, you did what?"

"I snogged your boss."

"That's disturbing," he said, smiling regretfully.

Logan pops his head into the office, while the autoclave makes a racket behind him. "Eh, I know it is not my business, but you are not going out with those two maths lecturers, right?"

"Noooo!" Lucy and I lie simultaneously and Logan sits down.

"How old is he?" Lucy asks.

"I don't know, I guess about ten years older, maybe more…"

"So he will listen to a full cricket match on the radio, wearing his first set of orthopaedic shoes while you're in the full bloom of life?!"

"I wouldn't call this PhD experience the blooming part of my life. I can't. There *must* be better to come, there just *must*… Please Lucy, come to the party?"

"Hm. One bottle of wine between the two of us before we leave. Deal?"

"Yip."

"Poor guys," Logan chips in.

He knows Lucy is a lightweight and will be somewhat tipsy before we even get there.

Lucy goes into the lab and I start to write a message to Alex and Chris, to say we are coming to the Oxford Alumni Party. I am distracted by an email just in from Mark: a new research idea. It has a paper attached, from a large research lab in Canada that did something similar to what I should apparently do now. I notice the Canadian group had at least six people working on the project and in our lab it would just be me. Plus, I only have a year left. I write:

Hi Mark,

Thanks for the idea.

I am not going to do it. Period.

I have enough to do.

Best, Karin

After sending the message, my hands start to shake. *Did I really just do that?*

I am nervous but I can feel my chest bulge with confidence and pride. I know Mark is sitting in his office, with his email box open in front of him. It will only be a matter of minutes – perhaps seconds – till he reads it and storms in screaming. I take my cigarettes out of the drawer and hurry out. The moment I walk down the stairs I hear an office door slamming.

I know it's him, but I don't turn round. I am too far down the stairs, he hasn't seen me, I think. I sit between the chemical containers and smoke a cigarette, overhearing Felix's colleague, Simon, telling a girl on the other side of the container about every aspect of his single-ball genitals. I listen to this bizarre monologue for a while, to distract myself from what might happen upon my return. *He will just shout at you. It's just words… just words… just words…*

Slowly I walk back upstairs and open the door of Lab 262. Lucy is standing at the bench, looking neutral, paralysed, silent – while Mark is red-faced and bellowing at her. He sees me come in and shoots a fiery glare at me, but he keeps his focus on Lucy. This is very unusual. Lucy is not a target. She is not the kind of person to attack or be attacked. Except for her, all Lab 262 inmates – even Barry – retort in some way when Mark verbally assaults us. It might be a weak retort issued purely for form's sake, but at least we respond. Lucy does not. She stands as if nailed to the ground and swallows every sentence quietly, keeping her large green eyes on Mark. She looks extraordinarily sad. It is touching and horrible and surely unnerving for Mark. I hurry to the freezer in the small side room, take some random samples out and pretend to work. The rant continues for another minute or so and then he closes on me. I know I am next. He ticks his keys on the bench next to where I am working. "You have enough to do?" he asks with a quite calm voice, but shaking his head.

"I have," I say, as firmly as I can.

"Fine," he states, and storms out of the lab.

Wow!

The new PhD student, Linn, comes out of the office into the lab as soon as the heavy lab door falls close. "What was that all about?" she asks.

"I haven't got a clue," Lucy says shrugging her shoulders.

Nodding at me, she adds, "He was asking where you were and when he couldn't see you he shouted at me instead."

"Yeah, I sent him an email… then I hurried out before 'Mark the Molotov cocktail' could fly in."

"What did you write?" Linn asks, very curious.

"I objected to another new research idea he proposed, by writing 'I am not going to do it. Period.'"

"Could you please let me know next time when you are planning on sending such a thing? So we can *all* pack up and leave… for a week," Lucy says, sounding bruised and disgruntled.

"Or we deploy your email replies to Mark to bully project students we dislike," Linn suggests. "We all make our escape and leave them behind!"

She is smiling; it's not a serious suggestion but it is seriously dismissive of Mark. Now here is a girl who has learned the ropes fast. "That's the spirit!" I say.

"I need a drink," Lucy says, in a defeated voice.

"It's three o'clock."

"Who cares! You're coming!"

"Of course."

Divide and rule is, of course, the *modus operandi* of all tin-pot thugs since time out of mind. But Mark is on a hiding to nothing. Lucy and I will huddle in the trenches together. She will stop blaming me for Mark's attack – she already did the moment I agreed to help her flush down the frustration at a bar. And I will despise him unto death for picking on Lucy.

Chapter 32

The Oxford University Alumni Party is taking place in the Dynamic Earth Museum, on the edge of the Old Town, close to Holyrood Palace. The doorman smiles kindly while opening the glass door. I am struck by his confident and serene style. He doesn't look like a man who would ever receive an invitation to the Oxford University Alumni Party, but he does look like a middle aged guy who has made peace with a tough life.

The corridor leading to the reception room is blocked by a group of people who all seem to know each other from their time at Oxford, or at least from the annual Alumni parties in Edinburgh. There is a table next to the door with small, half-filled glasses of champagne – "welcome drinks."

"I hope that's not all they have!" I whisper to Lucy much louder than intended.

"I guess they do refills?" Lucy doesn't bother trying to whisper.

It is seven in the evening and Lucy is midnight drunk.

As promised, we shared a bottle of wine before heading out to the party. Alas, Lucy drank the lion's share. We had left the lab one and a half hours earlier, bought the wine on the way to Lucy's flat, changed into fancy clothes, downed the wine and walked to the pub beside the Museum to meet up with Alex and Chris so that we could arrive together. At no point did we eat a morsel.

"If you are going to snog this Alex all evening I'm going home!" Lucy had warned me before entering the pub.

"Don't worry, Lucy. I don't think the Oxford University Alumni Party is the sort of place where one does that."

© Springer International Publishing AG 2017
K. Bodewits, *You Must Be Very Intelligent*,
DOI 10.1007/978-3-319-59321-0_32

Alex and Chris were both sitting at the bar, with a glass of spirits. Alex made eye contact with me instantly and I felt myself blushing. *Have I regressed ten years?!*

"Hi," he said, smiling warmly.

"Hi ladies," said Chris more excitedly.

"I bet they learned that during a soft-skills seminar," Lucy whispered in my ear, not very subtly.

"Double Vodka, girls?" Chris asked. *How charming... And how appealing... except that Lucy is already incapable of negotiating road traffic...*

"Make it a single," I said.

We chit-chatted about working at the School of Chemistry, but we avoided talking about Mark. Afterwards, we crossed the street with me keeping a wary eye out for Lucy amid the traffic.

We wade through this remarkably crowded museum corridor, grab a welcome drink and walk through a high door and into a white, sterile and way-too-bright reception room. Even though the room is full of interacting people, there is barely any noise. It's almost ghostly, as if the rocks of Arthur's Seat, which loom massively in the large windows, are absorbing all sound.

But there is simply no sound. This is an occasion during which you are expected not to talk at all, or only in the softest and most self-conscious tone – like you do in a church. The people around us look stiff, most unparty-like.

The place vividly calls to mind one of the posh restaurants Daniel's parents had taken me to a few years before. It too had a memorably white glare and was eerily quiet. No prices were displayed or mentioned. I guess money was no object to their target market but I couldn't help wondering what an evening like this – wherein you seem to bite your lip as much as the food – would cost. All three dishes on the menu had sounded alien to me, and deeply unappealing. I had finally picked the safest bet, a fish grilled in a layer of mud. I had dutifully toiled through the meal. By the time I had finished it I felt like a two-year-old after an open-mouthed full-face dive in the sandbox. In the car on the way home I had tried my best to get rid of the grains of sand between my teeth while Daniel's family discussed how wonderful the restaurant had been. It was one of those evenings when I was happy to be a misfit in Daniel's family.

Tonight is just as posh, with Alex and Chris both wearing smart suits. Both of them are soon inveigled into *sotto voce* conversations with old friends. I am standing with Lucy, whose alcohol intake is alarmingly obvious in her posture.

"I need to go to the loo, please behave, okay?" I say.

"Of course," she says and laughs loudly as if this is the most ridiculous request ever.

The noise of her laughter bulldozes through the silence of the room. Numerous eyes turn to us, less than approvingly, as if laughter at a party suggested a dubious presence. *Yikes, what a bunch of stiffs!*

"They're all watching you now," I say to Lucy, and head to the bathroom.

On my return I notice the time on the huge design clock in the middle of the room; nearly eight but my stomach says it's late evening – it hasn't seen food since eleven this morning. My body starts to send out alarm signals: disorientation, sweating and cramps in my belly. I know too well that I'm suffering from low blood sugar – it happens often these days – and it kicks in sharp with no warning at all. A few weeks ago I felt a murderous rage towards a waiter at a bar when my sugar level dropped. It had taken him over ten minutes to bring me a packet of crisps – hardly a capital crime. Vlad had joked nervously with the waiter, "If I were you, I would bring it to her now. Otherwise I'm afraid she will do something ugly to either you or me."

I know I need something to eat very quickly or I will keel over from the dizziness.

I scan the room for Lucy. I was only away for two minutes but inevitably she has been captured, by a handsome young guy. I might have taken him home without question, but Lucy looks utterly bored.

"I need food," I say, pulling on her arm and cursorily acknowledging the dude.

"Okay."

"I will bring her back to you later," I say to Mr. Gorgeous.

"Oh, please don't. He is boring!" Lucy says way too loud with her back towards him.

"She's not coming back, I'm afraid," I say.

"Got that," Mr. Gorgeous says.

I push her to the single table at the back of the room, next to a door leading to a kitchen; it has canapés…

"You think we can eat that?" Lucy asks and looks to me, only to see that I am already snacking.

"Why not?"

"Looks like it is for the waitress to pick up and take round the room."

"Possibly. But she'll live, and I feel like I won't."

I shove pastry after pastry down my throat. Ladylike, it is not. Lucy starts to eat too, but in a much more gentile style. I don't even taste the food, I just cram it in like fuel – which, right now, it is for me.

After we finish at least a quarter of the food on the table, Chris approaches us.

"Are you girls okay?" he enquires, looking concerned at the plate we just sort of raped.

"Just getting the most out of the entrance fee you paid for us!" I smile.

He's brought the louche common girls into the inner sanctum, but he can't suppress a smile. "I see."

"Is there really nothing more to drink than the tiny glasses of champagne at the entrance?" Lucy asks in a tone suggesting Chris go and remedy the drinks deficit right now.

She is holding a piece of bread with cheese in one hand, and toast with caviar in the other. She looks like Lady Muck gone right.

"I'm not sure," Chris says.

He seems to realise he has brought us to what we think of as the dullest party ever. "Are you not thirsty?" Lucy blurts out.

"I'll find a waiter and see if we can get something more to drink."

He goes off to forage for alcohol and I say, "Let's get away from here, before we get dirty looks for eating all this food."

Just as Chris is returning with two fresh glasses of champagne, a female voice fills the room, demanding everyone's attention. When we all obediently line up in front of her she informs us – in Wikipediot style – about the history of the Oxford University and how excellently it is performing in international university rankings.

The whole rankings thing pushes my buttons these days: I don't believe in any sort of rankings anymore, not since the start of my PhD – it's all bare-faced whitewashing, marketing drivel and damnable statistics. Like Oxford, the University of Edinburgh is ranking incredibly high internationally; even the chemistry department is doing well. Often as not, it means zip.

Lucy signals towards the back of the room. Two waiters are engaged in a frantic discussion about a half-eaten plate of food. I try to suppress my laughter but it escapes as a loud snorting noise. Again, Lucy and I are the focus of attention for a few seconds.

The lady holding forth with the microphone tells us how much Oxford University relies on donations from the alumni network. For half an hour she explains very precisely all the possible ways in which we can donate money to Oxford University. She closes with the assurance that she will still be here until the end of the party, which is in half an hour or so, should anyone have any question regarding the fascinating subject of funding for Oxford University. And she wishes us a lovely evening.

"It's finishing in half an hour?" Lucy is incredulous.

"Apparently so," I say.

"But we have been here less than an hour! What did they pay for our tickets?"

"Don't know. Twenty pounds? It's a very hardnosed fundraiser, and bloody far from the leisurely cognac-drinking romantic time I wanted!"

Lucy pauses for thought, then smiles, "Good we ate all the snacks!"

UNIVERSITY ALUMNI PARTY

„ Oh yes, in there…"

Slowly, the people spread over the room again and have quiet, superficial chats. Alex is standing far away from us and, for the third time this evening, is being approached by a particular middle aged female with reddish hair. The lady is gesturing wildly, obviously telling him something that excites her. I am too far away to overhear what she is saying and I cannot discern if Alex is greatly enjoying the conversation or is supremely bored by it. I wonder how he feels being so attractive to so many different women spanning generations? Maybe it's like sharing a bed with one girl one day and her mother the next? I'm not actually sure if he shares his bed with middle aged women. Somehow I suspect he does.

There is an old man leaning on a stick talking at Lucy. His voice is loud, he wears hearing aids. "In which department did you study?"

"Oh, I haven't been to Oxford," says Lucy cheerfully, sipping Champagne. "Those two guys over there brought us along."

He looks confused by the answer. "We are PhD students at the University of Edinburgh but those guys have studied at Oxford," I add.

"So what kind of job do you do?" Lucy asks.

"What did you just say?"

The old man points at his ears to indicate his hearing difficulties.

"What kind of job do you do?" Lucy repeats at a volume the whole room can pick up.

"I am retired!"

"Oh, really?" Lucy says, very loudly again, and as if this were beguiling stuff.

I whisper to Lucy, "Of course he's retired. He is at least ninety and has eyebrows that look like cat whiskers. Only Popes are still active at this age."

"I didn't notice. Shall we check the kitchen for more Champagne?"

Chris walks quickly towards us. "I think it's time to go somewhere else," he says.

You don't want us to go into the kitchen, do you?

"Have you donated money yet?" I ask playfully.

"I am donating my life to the academic system, I think that's enough."

We walk towards the exit and wait for Alex to finish his conversation. The middle-aged lady is walking him out. She hands over her telephone number with the words, "We live so close by, let's meet up soon."

She casts a disapproving look towards Lucy and me, smiles at Alex and walks back inside.

"She desperately wants to get fucked," I whisper in Lucy's ear.

"Yup. But she won't get it. This is not a place to pull. Trying it in here just makes you look desperate."

"I bet he'll meet her in a few days."

"You care?"

"No."

The four of us set off for The Voodoo Rooms, an unashamedly decorative night club in the New Town; escaping the paltry party of money-grubbing, modern academia for old-fashioned grandeur and devil-may-care fun.

Chapter 33

Around me, I see a tidy room featuring a perfectly made bed and a closed wooden wardrobe. There is a small bedside table sporting one book neatly placed on the top left corner of it. I presume the inhabitant is currently reading it. On the other side of the table is a perfectly ordered bookshelf. Various items, including a bottle of cologne, Scottish whisky and a baseball glove sit on top. It is obvious this man lives in solitude. There is not the faintest whiff of an intrusion from a partner and any traces of short-term liaisons have been meticulously erased.

He stands behind me and grabs my hips. He kisses my cheek bone and I turn my head so our lips can reach each other. With his hands he slightly presses my jaws, forcing my mouth to open and let his tongue slide in. He pulls softly on my ponytail and shivers go through my belly. He is in charge of me. Superior. This is a whole new ballgame to me. I'm the girl; normally I dictate the rules of engagement. But due to the age difference this feels natural. He takes the lead and I follow. *And God, this guy is triggering.*

He turns me around so I face him and abruptly takes a step backwards. "You can stay if you want, but nothing can happen," he says. "I can't risk my career."

"Okay," I say, not really understanding why I would be a risk to his career. *What kind of sexually transmitted disease do you have that stops you from wanting intercourse?*

I feel disappointed about his statement, and the abrupt termination of intimacy. At the same time, I am drunk; lying next to someone for a change

© Springer International Publishing AG 2017
K. Bodewits, *You Must Be Very Intelligent*,
DOI 10.1007/978-3-319-59321-0_33

would be lovely, whereas performing might be a bit of an effort anyway – excess alcohol doesn't put anyone at the top of their game, though they often think otherwise at the time...

He undresses to his boxer shorts, walks to the wardrobe and opens it. The clothes within are perfectly arranged, of course. He takes out a T-shirt and hands it to me. "Here, you can wear that."

He goes to the bathroom and I take the opportunity to change into his T-shirt. It is far too large for me and far from sexy, but it feels oddly nice to be in his clothes. I hear the toilet flushing and he returns to bed. He presses himself up against my back. "You've got such soft legs," he whispers and playfully kisses below my ear.

I try to turn around to face him but he embraces me with his strong arms.

I can't help wondering if intimacy and tension will build up between us while we sleep in the same bed, his body wrapped around mine, and if this will give rise to uncontrollable desire and sexual abandon.

It doesn't. But it was a nice thought to fall asleep on.

When I wake up, a breeze is wafting through the open window. It cools my face and chest, but it is still far too warm in the bed. Alex is as close to me as when we fell asleep. I am terribly thirsty and I suspect there will be a smell from my mouth suggesting a rat died in it some weeks ago. However, I am surprised to feel more at peace than I have felt at any time during the previous year. If I had the choice, and if he could turn his body temperature down by about three degrees, then I would lie here in his arms forever.

The T-shirt on my back feels wet from sweat, I don't know if it's his or mine, probably a mixture. I try to establish more physical space between us without waking him up, but I do not succeed. He moves his head up and kisses me on my cheekbone. He checks his mobile for the time and, with sudden impatience, springs from the bed and strides to the bathroom.

"Sorry. Duty calling. In just seventy-three minutes from now there will be fifty students sitting in a lecture theatre waiting for me."

"Has semester started already?"

"Yep."

Within seconds he is showering.

"I've been enrolling new students since Tuesday and started lecturing on Wednesday, meaning summer break is passé," he says, not looking overly excited at the toast with melted butter on his breakfast plate.

"How are the freshies?" I ask.

I don't mention that Lucy and I have already met several of them, in an underground pub. We had known there would be a party taking place for new undergraduates of the University of Edinburgh, and we were keen to find out if the guys would still hit on us, or if we were now too old for teens. I think it was only partly a search for reassurance about our market value; I think we were also both quite enthusiastic about the idea of taking a young energetic guy home. There were plenty of post-spotty cute boys at the party, and some had nice holiday tans; a rarity among poor PhD students. Others had obviously spent years optimising their biceps and toning their virile young bodies. No doubt about it, there were hotties among them. However, after several conversations featuring revelations such as "my mum is a great cook" and "my parents are such great company," their allure dwindled to zero. Slightly disorientated, Lucy and I had gone back above ground.

"It's the usual mixed bag: some excellent students, some who can't read letters or emails, many who cannot write properly, some who can't wake up in time for their first ever lecture, and some perennials deluded about their abilities and just basically circling the drain," Alex says.

"Had any mums on the phone spelling out the learning difficulties and dietary wishes of their beloved and talented offspring?"

Mark told us he had been ear-bashed by just such a mum on the phone last semester. She had taken issue with the marking, and was adamant her son should have been scored higher. Where she acquired her in-depth knowledge of biochemistry was a moot point which she didn't care to broach.

"Not yet, but it will come. The high season is around the first exams. It's getting worse every year."

"Did your mum phone the university when you were an undergrad?"

He glares momentarily but smiles, realising I'm joking. The helicoptering phenomenon is new but expanding at a terrifying rate.

"Sorry, need to go, help yourself," he says, gets into his jacket and closes the door behind him.

I am alone in his flat. Still wearing his T-shirt, I walk to the small kitchen-cum-living room. Though I'm pretty sure he showed me around yesterday evening, it feels like I'm entering the room for the first time. Like his bedroom, it is all very tidy, rather mysterious, not deathly, but as if there is no active presence inhabiting this place; as if the person living here has been beamed away via a time machine. A red squirrel sits on the outside window sill eating nuts, which Alex presumably left there for it.

The kitchen is sterile, all food and cutlery neatly packed away in the different cupboards. The only thing standing on the working top is an espresso machine with two small clean cups next to it. That it's two cups is the only thing to suggest anyone else ever visits this space.

I stretch out leisurely while one of the cups fills with coffee. I see the squirrel observing my every move from the other side of the window. The previous evening was playing on my mind. It had started with the silly Oxford University Alumni Party and had vastly improved when we arrived in The Voodoo Rooms. I soon matched Lucy for ludicrous alcohol intake and badly wanted to dance. Alex and Chris clearly had no interest in dancing, and hung by the bar.

"Is this Chris not something for you?" I had asked Lucy during one of our smokes outside the pub.

"Are you kidding me?"

"What's wrong with him?"

"Pretty much everything! He just told me I should drop by his flat to check out his stuffed dogs. Apparently he is collecting them. After Dr. Wilson's virtual cat..."

"Okay, he's out."

We had chatted the whole evening, though less than six hours later I can't recall any of the topics. It doesn't matter. It definitely had been one of the better nights in recent history. The tiny clock on the table next to the sofa is telling me that it is already past nine. Just to be a bit rebellious, I place the empty coffee cup in the middle of the working top. I'm sure it will disturb him when he comes home later. *My chaotic personality and his pedantic self are such a misfit.* Very slowly I undress, savouring memories of the previous night. I get into the shower and wash myself with his soap. I like the sensuality but then I find it a bit disturbing to smell of shower gel for men when I am out of the shower. For a short moment I consider putting his T-shirt on again, but after lifting it from the bathroom floor I realise it smells of Bell's Whisky, testosterone, sweat and... *There's no way I can have it near my freshly washed body...*

My head starts pounding as soon as my eyes make contact with broad daylight. I slowly walk to my flat to change into jeans and sneakers. I pack some sports clothes, in case I get the chance to run today, and cycle to the university. A few feet after the last set of traffic lights before campus, a police car abruptly cuts me off the road. I have to stop. Two tall policemen get out of the car.

"What made you cycle through red lights?"

"Did I cycle through red?" I say, truly not sure if I did or not.

"You don't know?"

"Not really, no."

"We've followed you for the last few minutes. You cycled through *five* red lights in a row!" the more sympathetic looking guy states. *Yikes, what happens here?*

"Five?"

"Yes, *five.*"

"Is there a problem with that?"

Both policemen look surprised by my question. The less sympathetic one is clearly annoyed and takes out a block of paper ready to write down my personal details and launch into dreary legal shenanigans. But the other policeman steps in front of him and gestures to hold fire, for now.

"You are not allowed to cycle through red lights. You know that, right?"

"Not really, no. I thought the traffic lights were just for cars."

"Where are you from?"

"The Netherlands."

"In the Netherlands you are allowed to cycle through red lights?"

"Yes, of course," I say, as if this is the most normal thing on the planet and I have never heard anything else.

"We had a Dutch girl before, saying exactly the same." *Good girl!*

"Are you studying here?" he asks, pointing at the university buildings next to us.

"Yes, I am."

"How long have you been here?"

"A couple of weeks."

I know my story won't stand up if I tell them I just started the third year of my PhD.

"We should really tell the university to give a list of traffic rules to foreign students. What department are you in?"

"Chemistry."

"PhD?"

"Yes."

"Wow." He looks at me with deep respect. *Oh, if you but knew, kindly, innocent policeman…* "We should go and talk to them."

He said that a bit too enthusiastically for my liking. I desperately hope he does not come up with the idea that I take them to the university. Mark would definitely not be pleased to see a policeman standing in his office, and I strongly doubt he would confirm my claim that I just started my PhD – though he would happily add a couple of years to my sentence if he could.

"I'm terribly sorry; next time I will wait until it's green," I say, hopefully ending the conversation.

"Off you go," he says and even gives me a friendly wink. I get on my bike and cycle the last bit.

I ignore my headache and walk into the lab feeling full of positive energy. Mark is standing at the bench, talking to Linn. I feel Mark's eyes peering at me as I pass the benches on my way to the office. Just before I reach the office door he addresses me.

"You missed Barry's talk!"

He is clearly boiling with anger. *Shit. The departmental talk…*

I feel the morning's happiness fade to familiar nothingness as I turn towards him.

"Sorry, I forgot about it."

"You forgot about it?"

He obviously doesn't believe me, shaking his head and leaning on the bench with his hand. "Where have you been?"

"At home."

"Where is Lucy?"

"I don't know."

I wait for a few seconds, but it seems that neither of us has anything to add to the conversation, so I turn round to enter the office. Quickly I drop off my bag and coat, grab a Styrofoam container from the sink and head out of the lab to fetch ice. I'm not actually sure if I need any ice for my experiments today, but it is good to have some physical distance from Mark.

At a slow pace I walk over the glass bridge to the new side of the building, hoping that Mark is safely ensconced in his office by the time I return. For a few minutes I watch the machine spitting out flakes of ice. I fill up my bucket and, again, slowly make my way back to the lab. The sun is not far from the top of the sky and I am thinking about the evening before. My body is re-filling with positive energy as it had done this morning. *I stayed over at the hottest lecturer in the university.* With a year to go, all I can hope for is distraction. The possibility of sex tantalises regularly, but today it's weird to think that sometimes it is sex we didn't actually have which raises the spirits and keeps us soldiering on.

It is helping me get through this PhD, knowing that there is more to life than *this*. And *this* actually doesn't matter in reality; just going through the motions to collect a title. Normally I don't have the spiritual strength to be so depressingly honest…

Chapter 34

My Nokia telephone beeps from somewhere deep in my dreams and as I slowly gain consciousness I strongly suspect it has been beeping for a very long time. I reach out to the pillow next to me; since Daniel left Edinburgh it mostly hosts my mobile, often accompanied by my laptop. My eyes feel heavy and dry when I finally open them to check the time. 9:30 a.m. *Shit, two hours later than I wanted to get up.*

When I wrench my body from the bed I consider my options, and lying back down is oh-so tempting. As time passes studying for my PhD, it becomes harder and harder to get out of bed. It has been a long time since I've been first to arrive in the lab in the morning. Apart from a brute desire to get out of the lab as soon as my obligatory three years are over, I appear bereft of motivation.

In recent weeks I have unsuccessfully tried to ignore the fact that this PhD is *the* big thing in my life. It all seems so long ago, in a different life, that I studied hard for my bachelor's and master's degrees, to graduate with good marks; when I studied so much and got complimented for the work I was doing, which I found fascinating.

There is a pile of clothes next to the bed. I pull out the jeans and a bra, and throw the rest in the washing basket. Slowly I walk to the bathroom to brush my teeth. I look in the mirror and wipe some cold water around my eyes before making my lashes darker with mascara and picking a top to wear. I feel like I am putting on protective armour.

The only person in the office is Logan, sitting behind one of the shared computers.

© Springer International Publishing AG 2017
K. Bodewits, *You Must Be* Very *Intelligent*,
DOI 10.1007/978-3-319-59321-0_34

"Good morning," I say, and Logan turns in my direction.

"Morning, Ka. You look great!"

"Thanks for the sarcasm."

"You went to the pub?"

"Yep, with Lucy."

"How much did you have?" He smiles slightly.

"Oh I don't know. Isn't that a bit of a guy thing to count them, so you can bore all your friends with numbers later?"

"Beers?"

"Yes, dude. You know Lucy and I can't afford spirits or cocktails or whatever bullshit mankind came up with. We are not as posh as you, my dear."

Logan laughs. It has become a running gag to place Logan in the posh box after finding out that he was receiving a slightly higher salary from his industry funding body than the academic survival package the other PhD students in Lab 262 receive.

I walk to the lab, Logan following behind me, and get into my lab coat. "How about some music?" I ask.

"Sure. As long as it isn't radio and I have to listen to the same news bulletin repeatedly."

"Some people in the lab might benefit from a bit of news," I say, presuming that Logan would understand the remark.

Very recently Barry had announced to Mark that he would be buying a house nearby, with Babette. I had overheard the conversation and the information had baffled me. I have such a powerful aversion to both Babette and Barry that I don't understand why they do not have an aversion to each other. Then again some people also think a pit bull in the house is fun. A few wild cards out there, not overly concerned about losing fingers, are crazy enough to try taming honey badgers. So, why not Babette or Barry on your sofa?

"Is it a good idea to buy a house now?" I asked Barry as soon as Mark left the lab.

"Why would it not be?" he replied from the other side of the bench, scanning warily for Babette.

"Did you hear that the housing market is collapsing? And it might take years to recover, cheap to buy maybe but perhaps a poor investment too…"

Barry looked at me blankly.

"Economic crisis?" I tried, but still it did not seem to ring a bell.

"You do know that the economy is in tatters throughout Europe, right?" I asked this carefully while the eyes of the project student behind him slowly widened in disbelief.

"No," Barry replied, clearly being serious. Barry doesn't joke.

"WHAT!!?? Where the fuck have you been these past *three* years?"

"I don't read news." *You are a doctor!!?? Who gave you your degree?* The project student's eyes widened even more and silently he started to mumble "what?"

"You don't even need to read it! It's on the front page of *every* newspaper. It's all over the web, it's everywhere! The biggest economic crisis since the 1920s!" I might have added that this helped trigger the Second World War, but refrained from it as I didn't feel like explaining what a World War is.

"As I said, I don't read the news, not even headlines. It makes me depressed." *You are already depressed; what news could be more depressing than Babette in your bed?*

"Well, news is not like a humorous fairy-tale with a happy end every time, indeed... But you absorb it anyway, even just via comedians, no?"

"I don't like comedy." *Of course. Silly me.*

"What about Babette?"

"She also doesn't read news. It upsets her." *People saying hello infuriate her. I'd love to see her react to a global crisis.*

"In that case it might be the right time for you to buy a house," I conclude as if there is logic in my complete non sequitur. *Maybe the economy will actually benefit from clueless morons spending cash...*

Barry walked off. They bought the house, which is their business and good luck to them. However, they subsequently did something truly shocking: they invited all Lab 262 inmates for a housewarming lunch. It is a perfectly tortuous prospect, especially since lunch is traditionally when we get away from the hideous dynamic of our cell – we need that respite to get through the day. *This isn't friendly overtures, this is capricious stupidity. Pleasant relations between us and the KBL gang are long dead in the water, sunk without trace, dissolved in dank ocean depths; the mere idea of trying to dredge up bonhomie is excruciating...*

I search through the stack of CDs lying next to the stereo. Felix had given me a few old ones from the large CD collection in the Homer lab.

"*Radiohead?*"

"If it has to be."

I place a glass with water under the flame and await it boiling before placing my samples in it.

"Looking forward to the lunch?" Logan asks in a sarcastic tone.

"Aren't we all?..."

Logan grins. Thom Yorke is singing *Creep* in the background. "Feels a bit like a Little Red Riding Hood experience… not knowing if the wolf will eat us," I add.

The door of the lab opens quickly and the jingle of the keychain confirms my worst fear: Mark. Quickly I take my samples out of the boiling water so I can leave the lab. With only Logan and me left in the lab I know my chances of escape are low, but it's worth a shot. Mark moans, turns off the stereo and indicates I am to stop, just as I open the door. I sigh inwardly and let the door fall closed. I try to force a smile, I really do, but I know damn well that tension will be clearly visible in my unreal expression.

"You won the prize Karin, well done!" Mark says, happily.

I'm thrown. *Is he being genuine? Dare I believe he is happy with me? I don't know about any prize…*

"Which prize?" I ask resting my hands in the back pockets of my jeans.

"The best grant proposal award. You came first amongst fifteen."

Now he sounds annoyed, so it's real. He is annoyed I did not instantly know what he was talking about.

"Really?" I ask, excitedly.

"Yes. You won the flight to go to Toronto. You need to ask James to pay the conference fee for you." *Of course, you don't have a penny left.*

"Cool!" I say happily.

"Brian is going as well, so you can travel together." *Right…*

When he leaves I sing more than speak, "Canada, here I come."

It is not that I am particularly interested in the conference, though of course I might meet some good people, but whether I do or not, I am flying away from Lab 262 for a few days – I'm on cloud nine already…

I walk downstairs to tell Felix the exciting news and score myself a coffee. I enter the cubicle where Elizabeth is sitting, whistling her endlessly annoying tune. She holds a monkey hand puppet which normally sits next to her computer, letting it play with a few tomatoes on the desk. I suspect the tomatoes belong to William; he eats three tomatoes and exactly eight slices of rye bread for lunch – every day without fail.

"Good you're here, flower, I need a coffee too," Felix says.

"I get to go to Toronto," I say excitedly and tell him about the competition I won.

"I'm jealous! Homer doesn't let us go to conferences, unless we publish a world class paper first," Felix says.

He knocks on the window to the lab to indicate to William, standing at his fume hood, that we need the water tank. William takes off his lab clothes and approaches us with a smile.

"Hi Ka," he says friendly, but the moment he sees Elizabeth behind me playing with the puppet the smile leaves his face.

He walks towards her, rips the tomatoes out of her hands and throws them all in the bin. "Never ever touch my stuff again. Do you have a clue how dirty and contaminated this disgusting monkey is?" he shouts, pointing at Elizabeth.

"Sorry," she says neutrally and continues to whistle the all-too-familiar tune from her remarkably limited repertoire.

William joins us for coffee in the small kitchen and I'm struck that I wanted to celebrate my good news with inmates of another lab rather than my own.

I head back upstairs to prepare buffers, which I need for experiments in the hospital this afternoon. At midday Mark storms in.

"We will walk together to their new house and will all be ready to leave in fifteen minutes," says he, to Logan and me.

"Okay."

He checks the empty office and walks back to the lab. "Where is Lucy?" he barks, looking at me.

"She is at home, writing her thesis," I say.

"What do you mean?" he asks, implying I said something terribly wrong.

I wonder how I can elaborate on a self-evidently simple and comprehensible statement. "Eh, she is writing up her thesis," I repeat, trying to sound much more secure than I feel.

"She is not ready yet!"

He is properly shouting. Yet he knows Lucy's stipend ran out in summer, and he knows she is writing up her thesis. He knows fine and well she is not being paid to work in here – that is long understood.

"Um. She is actually moving into my living room this weekend, to save money," I dare to say, hopefully getting the message across that Lucy's life without a salary is tricky.

Mark sighs loudly – as if Lucy's predicament was *his* inconvenience – and storms out.

During the lunch in Babette's and Barry's small house, decorated in the style of a recently deceased curmudgeon, I notice the seating arrangement; the same as for lab meetings. Mark, Barry and Babette talk about KBL. Linn, Logan and myself are designated figurants as usual. I regret that I came, Lucy had been right. *What are we doing here?*

Sad fact is: I didn't dare to *not* show up. I pour Logan, Linn and myself a glass of wine from the bottles on the table. I take a few large sips, and Logan is in position to give me a re-fill before I see the bottom of my glass. I know that, with the alcohol in my body, my experiments in the hospital will have to wait till tomorrow, but I don't care. All I care about is making the here and now bearable. The alcohol does its job. I dutifully listen to the incessant prattle of Mark in a house of people I loathe. I'm not really here, of course. My mind is far away, in a world where people just might take a polite interest if I dared to speak.

Chapter 35

"I seriously don't even know the names of the different types of glassware which chemists use," I say to Logan, crossing the glass bridge to the new part of the chemistry building, to the teaching labs. Logan laughs. In the lab he has been mostly surrounded by females for the last one and a half years of his life. He has become accustomed to faint hysteria about stress.

"It's just redox reactions you're teaching. You'll be fine."

"That's my problem, I might as well teach ballet. I don't know anything about redox reactions!"

Logan laughs even louder. "You know more about them than the students know."

During the previous two semesters I taught biological chemistry to first year students. Perhaps that went a little too well because this time I've been signed up, by Mark, to teach chemistry to second year students instead. Following my first year exam, I put in an effort to get to grips with chemistry but if this qualifies me as a teacher then God help the future of chemistry. I am, after all, a biologist. Rationally, I know I probably have nothing to worry about; undergrads don't ask much. But even so, I fear some overly enthusiastic student could shoot me down in flames.

We manoeuver through the hallway, between groups of students sat on the floor. They are waiting for the class to start. As soon as I see them I feel less nervous. None of them looks terribly motivated and the course

© Springer International Publishing AG 2017
K. Bodewits, *You Must Be Very Intelligent*,
DOI 10.1007/978-3-319-59321-0_35

manuals are peacefully sitting in colourful backpacks – if indeed they are even here.

"Why did Mark sign me up for this?" I ask before entering the large, new teaching labs.

"To bully you."

Logan is teaching at a bench near the front of the room and I am right at the back. The two ladies running the teaching labs give us a few more instructions before they open the large white doors and let the students in.

The students shuffle in slowly, as motivated as a herd being hounded to an abattoir. They search the lab quite leisurely, looking for the number on the bench which indicates where they are supposed to be taught. During the next few weeks they will circulate in groups of ten throughout the room, ensuring they partake in all the experiments. As PhD students in teaching roles, we will stay at the same bench running the same experiment every week for a full semester. Gratifying, it is not.

"What did they smoke?" whispers Rostek, a tall Polish guy, doing his PhD in the inorganic chemistry section.

"They probably had a party last night," I say.

"Yes, too much vodka."

It takes several minutes for the students to accomplish the apparently Herculean task of changing into lab coats and putting on safety specs. I explain in three sentences what we are going to do this afternoon and how long the experiment is supposed to take. Logan, who taught the experiment last semester, told me that my group should be finished first. That prospect pleases me enormously. As both the students and myself are keen to get this over as soon as possible, I encourage them to start pipetting the mixtures together straight away which, in fact, they do with commendable obedience.

While I wait till the students on my bench complete their work, I hear Rostek talking in an engaging manner about the experiments. Bursting with enthusiasm and passion, he writes chemical structures and experimental set-ups on the whiteboard as if there is nothing more interesting in this world. He is a born teacher, a born chemist. It is right and fitting that he is here. It is wrong and ill-fitting that I am here. I wish I could be as happy and passionate as Rostek, but this subject simply isn't my bag...

My students are making quick progress, as Logan predicted, and no one comes up with the idea to ask me about the theory behind what they are doing. The first couples are packing up to leave within two hours of us starting. Just as the last two are finishing, one of the ladies running the lab appears at my bench with two more students. Their eyes are red and they walk as if they just stepped off a boat after a long voyage in stormy waters.

"These two gentlemen are in your group," she announces.

"It's too late now, we're finished," I say, sort of knowing I probably won't get away with it; I will be penalised with an extra two hours of teaching, for this pair partying too hard to get to class on time. "Let them come back another time."

"Tomorrow the groups are full," she says, at length, as if presenting me with a problem I should care about.

"Maybe there is a spot free next year," I say, knowing the pair will never be sent back a class.

The students stand there with eyes wide open. It is obvious they are accustomed to being treated as valued customers at the university, and have never come across such outrageously shoddy service before.

"I'm afraid they have to do it now," says she, pointing at two empty seats on my bench before striding off. *Learning that it is fine to behave like useless brats; that's "education" in this charade...*

I sigh and regard the students with much derision, but they don't know or care. They change into lab coats slowly, like we have all the time in the world, and conduct the experiment. I'm sitting, waiting and looking around a bit. Rostek walks around very actively, answering all kinds of questions. There are two students standing at the fume hood, and Rostek decides they need help too. He takes the separating funnel filled with solvents out of one student's hand and says: "You really have to shake it properly with two hands; not just swaying, like you just did."

He holds the funnel at both ends and shakes it vigorously in all directions to mix the solvents. "Like this," he adds, before handing the glassware back to the undergrad.

"But Rostek," says the girl. "I only have one hand."

My eyes and Rosteks' eyes move down at the same time, me observing her from a more comfortable distance. The girl indeed only has one hand, the other arm stops just before her wrist. *What on Earth...*

Rostek stares at her in disbelief, not knowing what to say. I'm sure he is wondering how someone with only one hand can be sent to an organic chemistry lab class. He looks in my direction as if I might have a clue how to handle the situation. I shrug my shoulders and shake my head slightly. "What is happening to this world?" he says in a much stronger Eastern European accent than he used before. "That is quite dangerous," he adds, after a few seconds of silence during which he took the funnel back out of her hand.

There are audible traces of shock in his voice and manner, "What are you doing in a chemistry lab?"

The girl looks crushed, but anger creeps into her expression. "Ever heard of equality laws?" she squeaks as if she has been properly offended.

"Equal rights is all good and fine, but a blind person can't become a bus driver. And in my opinion someone with only one hand can't go into a lab without help. Theoretical chemistry, yes, but enrolling for a strongly lab-based chemistry degree is not only endangering yourself but everyone around you. Did nobody tell you this is not a good idea?"

"No," she says upset, but she sits back in her chair and awaits her study buddy finishing their experiment.

Around 5:00 p.m. all students, apart from mine, have finished. I am still sitting at the bench, with Loser One and Loser Two, when Mark walks in to fetch the attendance sheet. When he sees me sitting at the bench, he walks towards me. He is smiling in a jolly way, as he usually does when there are "outsiders" around. I haven't seen him for over two weeks; I had been in Canada and then he had been at a conference in Turkey.

"Hi, how are things going?" he asks enthusiastically.

"Fine, those guys just decided to sleep in so we aren't finished yet," I say, pointing at Loser One and Loser Two.

"Hmmm, next time you come in time, will you? You are wasting the time of my PhD students."

Mark speaks in a tone that brokers no objections. They nod and continue the experiment like two beaten dogs. *Wow.*

"How was the Toronto conference?"

"Yeah, I enjoyed it, thanks. Weren't many people there; about sixty I think."

At the start of the conference I had hooked up with two girls I didn't know before, both late stage PhD students, and we had a good time together. None of us was overly fussed about attending the sessions. We were there for a break, from daily lab routines and our dragging PhD projects. One of the girls got upset when she saw her own data being presented by a different research group on the first conference day.

"That is my data," she whispered during the talk.

"Huh... how come they have it?" I asked.

"My boss gave it away in return for collaboration on a different project."

"What about you?"

"I wasn't asked, and now I have to stay on for another year to generate other data for my PhD thesis," she said, with watery eyes.

"Great boss."

"Yeah, he's a dick."

Tears were running down her cheeks. It was unexpected and sad, but an excellent excuse to skip the next session and spend the rest of the day in a pub instead. We partied hard and discovered all three of us were passionate scientists who are being demotivated by ropy PhD experiences. We had fun, and we realised we are not unusual. That's oddly comforting – yet distressing from a neutral point of view.

"You travelled with Brian?" Mark asks, leaning on the bench of the two undergrads. *Was too dangerous – we might inspire the other to jump out at 30,000 feet.*

Being at the same conference was one thing, but actually sitting next to him on a long flight was unthinkable. I don't tell Mark that Brian and I merely said hi in Toronto and thereafter cold shouldered each other until the flight home. We were on the same flight, and we dutifully sat together in the airport. And we both sighed with relief when we could finally board and take a seat far away from the other.

"No, unfortunately not. I took the opportunity to see the city and Niagara Falls, so I travelled two days earlier."

"You liked the Falls?" *No, this natural-wonder-gone-Disneyland sucked big time.*

"It was impressive," I say, wondering if Mark and I have ever before had such a normal conversation?

"Okay. Any news from the conference?"

"Well, the presentations were like in Venice, mainly research that has been published, but Prof. Clark and Prof. Green took their collaboration to the next level," I say, smiling.

Mark raises his eyebrows, clearly awaiting elaboration; Clark and Green are two of the most eminent researchers in the field of cystic fibrosis. "He banged her," I add.

Mark expresses a mixture of shock, confusion and worry as if I had just used a bad word that the little boys at the bench should never hear. "How do you know?" he whispers.

"With two other PhD students, I bumped into them. We came back from the pub, heard some strange noise coming out of a meeting room and had a look to see what was going on. And there they were; on top of the piano – a grand piano of course."

„ We are just collaborating…"

Mark looks entertained yet unsure how to react. Apart from some drunken comments during our Christmas night out, our lab is bereft of erotica when Mark is around. The guys at the bench definitely overhear the conversation, but pretend to be absorbed in the work they obviously prioritised lower than yet another beer in the pub the night before. "'We're just collaborating' Prof. Clark said when he saw us standing in the doorway. It was a plausible statement since he had both hands on her ass and she had lifted one leg onto the piano – they were jointly committed to the project, indeed."

"Okay, okay, enough!" *He must be visualising Prof. Green with one of her dwarf legs on the piano. I liked that I had furnished him with sufficient detail to do this.*

Prof. Green is small and has chubby legs that she likes to pack into a tight leather skirt. I had previously seen her at the conference in Venice. She is an excellent researcher, no doubt, but also the perfect personification of academia as a drip can of eccentric species. I love watching her; how she laughs, how she dresses like the hotty of the local flea market, how she endeavours to be sexy. I remember her pausing during a lengthy monologue at breakfast to dart to the buffet, grab a sausage soaking in fat and slide it into her mouth in one go. Prof. Clark, by contrast, is tall, handsome and well-dressed. He is also married. And he had proudly announced at the start of the conference that his wife had just given birth to his third daughter. They are the two professors who did, after some talks, get a good discussion going. However, the day after their piano sex scene they both sat through the sessions looking red-faced and ashamed.

Loser One and Loser Two are slowly packing up their stuff, and Mark and I follow them out, locking the door behind them. The guys leave the building through the nearest exit and Mark and I walk back to Lab 262.

"You happen to know where Lucy is hanging out?"

"She is still writing up, at home."

He shakes his head. "Tell her to come in! She is *not* finished yet."

"She is still working weekends," I say, not mentioning that she picked this time schedule to avoid Mark, who spends weekends with his girlfriend in Stirling.

We cross the glass bridge back to the old building. When we have almost reached the end of the bridge the first safety door swings open. A young handsome guy is walking toward us, smiling at me; it is Alex. I freeze in mid-motion. *What the hell is he doing here?* My mind works fast enough to realise how ridiculous I must look to both Mark and Alex, but my knees seem immobilised for a few seconds. Alex is a step closer and I can smell the triggering mixture of cologne and body odour. He lets his eyes roll over my body and I instantly regret wearing my jeans. I am looking at his face, but I can't stare at him too long because I am flustered and Mark is examining me. *Move! Move!*

"Hello," Alex says, unclear if he is greeting me or Mark or both.

"Hello," both Mark and I reply, and I tentatively take a few steps towards the door.

Mark is looking at me, puzzled, "You know Alex?"

"Not really. I know one of his PhD students."

After the Oxford Alumni Party we had met only once. Alex had proposed we spend an afternoon together in the seaside town of North Berwick, which we did. I had been excited to go there with him and left work around lunchtime. Alas, when we had finally strolled over the beach and climbed over rocks, I could barely talk. I felt such a need to impress him, find the right words, contrive the perfect message for the perfect man, but I was simply too drained of energy to do so. Though he still failed to mention it, I knew from both Chris and Greg he was dating someone else. This should have taken the pressure off me, but it didn't. He did his best, but his mere presence stifled me. I had stayed at his place another night, and we had a lively chat at the breakfast table about university policies, particularly around lectureships; how they become more and more like administrative jobs and forever move away from research, how the percentage of time he spends on each activity is monitored non-stop for everything he does, how alarmingly spoiled and spoon-fed students are becoming, and how tragically much time is invested in parenting them. Research became a "hobby" for your "leisure time," but one that you're appraised for, which translates into promotions. We had talked a bit about Mark, and Alex had confirmed he has a bad reputation at the university. After that day on the rocks of North Berwick beach we had some email contact, but that was it. I guess we both silently concluded that it would not work between us; no bad feeling, maybe in another life… but not now.

We walk through another set of doors and enter the dark corridor where Mark's office is located. We smell cigar smoke, permeating from Prof. Gilton's office. Occasionally, he smokes outside and we have a friendly chat between the chemical containers. Other days he doesn't feel like walking downstairs and we see the smoke billowing through the slit of his doorway. To so completely and utterly neglect the rules and policies in a place where you are supposed to be a role model to students is appealingly cavalier, at least to me. I came to like him in the last year. After my first year exam he enquired a few times how I was getting on and explained that the exam had been a "normal procedure." He seems to care deeply that the people in our lab can finish their PhDs. He hardly ever enters our lab, but he talks to the Lab 262 inmates in the corridor, offering his help whenever he can.

Mark unlocks his office door and I can only hope he is not inviting me in to continue our conversation.

"Did Brian tell you he found a lectureship position in Cork?" he asks, standing in the doorway.

"He didn't."

"You can join him to Cork for a few months to help him set up his lab." *Are you really this clueless?*

"I would like to finish my PhD first, have a nice evening," I say hurrying off to the lab, though just to fetch my coat and backpack.

It is still early but the office is dark and silent. Until two days ago the fossil autoclave had been hissing around this time of the day but an inspector had recently done some inspecting and – to absolutely nobody's surprise – closed the old beauty down. Now it just stands there, waiting to finally rest in the knacker's yard where it has long belonged. I lock all doors and cycle home.

I say hello to Lucy, who is sitting at the desk typing on her laptop. I sit on the sill, stare out of the window and light a cigarette. "Mark wants you to come in and do lab work," I say.

"I bet. He writes me emails all the time. I actually found a teaching position in Senegal today. I decided to take it. I'll write up till I go." *Noooo!...*

She is excited but I'm sad. I knew she was looking around for jobs to get away from Edinburgh. Of course sleeping on a mattress in my living room is nice for a slumber party but in the long run it is poor testament to a full life for a functioning adult. I just didn't think she would find something so quickly.

"I guess I should be happy for you. When are you leaving?"

"In a month. I still have about four weekends of work to finish and then I should be ready to leave."

"So soon?"

"You'll be fine. You have Felix… and William."

I know. I'm a big girl. But still, the best thing about my PhD is going to West Africa, and work looms large in life when empty spaces appear. I sometimes think Lucy is one of the main reasons – even *the* main reason – I do not suffer from depression during my PhD.

Chapter 36

Sunday morning 8:30, I open my eyes and stare straight into the back of a dude. *Yip, that is definitely a dude's back – in my bed. I presume it will have a dude attached to it... What the bloody hell is he doing here?! And who is he?...*

Image by blurry image, the night before comes into vague focus... And what had seemed like a fine and fun idea – to bring the Hungarian pizza baker home – now seems to have been inspired solely by alcohol. *Why did he not just leave? Hasn't he got some dough to knead? What is his name?...*

I hobble to the bathroom to empty my pressing bladder. My stay in the bathroom is longer than necessary as I take my time to torment myself with the events of the previous evening. The general narrative is clear and clichéd though lacunae abound, and my dignity does not appear to feature anywhere...

Lucy and I had been in Dragonfly, a bar in West Port, close to the Grassmarket. We had been sharing our initial PhD interview experiences – our first oh-so innocent encounters with Mark and the Lab 262 inmates. The story of Quasimodo wanting sex on the bread toaster bed in Pollock Halls came up, inevitably followed by the story of the good-looking guy I had met on the street that same fateful day. I had sentimentally kept the note with his number in my wallet ever since. On discovering this, Lucy had encouraged me to dig it out and ring him.

"Call him, invite him over," she said brimming with wine-fuelled largesse on the bar chair.

"Yeah, why not?" I thought, brimming with wine-fuelled derring-do on the seat beside her.

© Springer International Publishing AG 2017
K. Bodewits, *You Must Be Very Intelligent*,
DOI 10.1007/978-3-319-59321-0_36

A couple of glasses to the wise, it seemed a downright logical thing to do. I took out my mobile and rang the number of a total stranger. Explaining who I was – a girl you met very briefly on the street yonks ago – was a lengthy process. It wasn't a matter of convincing him that he had found me attractive; it was a matter of convincing him I was not mad, desperate or pathetically drunk, and it was all in the tone, never overtly stated. In the end he very happily joined us.

"No chalk on the windows with this one," I whispered when the guy, whose name still eluded us, went up to the bar to fetch another round of drinks.

"The IQ of a fish-finger," Lucy estimated, quite generously. "But he's hot!"

After that round, Lucy left and things got hazy. I know I did invite him home, into my bedroom, and that we had sex, but that is about it.

I feel disgusting and opt for a shower, a girlie attempt to wash away the reality. With shaking knees, I climb in the bathtub and warm water runs over my body, my face and hair. I manage to convince myself that with just one Ibuprofen, half a litre of coffee and another short nap, the day will be fine. *If pizza baker would just kindly exit stage left pronto!*...

Right now, a stealthy text to Lucy seems the best option. I wrap my aching body in a towel, get my mobile out of my coat and plump onto the toilet seat: "Lucy, there is a dude in my bed. Could you please get rid of him!"

As soon as I press "send" I hear the beeping noise of her mobile in the living room. With tired legs I waddle back to the bedroom, lie down next to the pizza baker and wait.

In the faint light coming through the curtains I can vaguely see the details of his back. A few lost hairs, some spots and some curious white bits and pieces. I survey this white matter for a while – I have nothing else to do – and conclude these could be skin flakes. However, taking the occupation of the Hungarian into account I wonder if they are actually dried-out pieces of dough. I could easily test this theory by wetting my finger and seeing if the dried flakes turn into something malleable. But I decide that would be rude and I don't want to touch him in case he interprets it as erotic desire. In that case I would have to come clean; "God God, no! I was just wondering if you had decaying foodstuff on your back."

The living room door opens as Lucy makes her way to the bathroom. At least it didn't take much to wake her. The sound waves from the toilet flushing temporarily fill the house, and are followed by clatter as Lucy moves heavy gear around. Her footsteps come close to the bedroom door but she doesn't enter. The light snore of the Hungarian pizza baker is unperturbed by all this. I have no idea what Lucy is doing but then the whole house fills with a terribly loud cock-crow followed by the Beatles song *Good Morning, Good Morning*. She must have pointed the large stereo boxes right at the bedroom door and put the music on full volume. Now she is banging on the door and shouting "Ka, get up! Get up! We have to go to church now!" *Church? Church, really? Do you think the poor guy is that stupid?...*

The pizza baker sits up straight in bed, totally overwhelmed by Lucy's ruse. He looks at me like he wants me to do something about the noise, but the only thing I can do is laugh and shout, "Well done, Lucy!"

He gets dressed, though a bit too slowly for my liking. Just before he finally leaves the bedroom, he says, "It was very nice and maybe I see you again?"

I don't want to be rude or cruel, but neither do I want him to imagine that he might ever rest that dough-splattered back in my bed again, "Well, I'm glad that one of us sees potential..."

The front door opens and closes a few seconds later. He is gone. I look at the clock, 9:00 a.m., still early. Lucy left just before the pizza baker, so I can have a little nap before going to the lab. I can wake up with the Hungarian memory consigned to the dustbin of history.

On Sundays it looks quiet on the King's Buildings campus. If you peer closely – through the windows of the labs – you will find that there are in fact many people working, mostly PhD students, some postdocs. A professor or lecturer might occasionally be spotted, probably just fetching something from their office which they meant to take home. What defines the scene is the lack of undergraduate students walking from the bus stop to the lecture theatres, filling up the teaching labs, having their lunch in the canteens or corridors. During term, this is a thriving university environment brimming with people who believe the world is at their feet and their parents' investment will soon lead to returns. At the weekend it is a joyless place, hosting mostly people that – at least once upon a youthful day – are longing to become real scientists. Many spend hours upon hours here at unsocial hours, desperately trying to salvage the treasured dream fading to grey amid forbidding cloisters – never to be revived.

The statue of Joseph Black looks sad in the late autumnal drizzle. I park my bike at the entrance, on top of a thick pack of yellow leaves from the majestic trees which adorn the pavement. When I walk upstairs I literally bump into Ajit, the unsporty, short Indian guy who works in the organic synthesis group next door, and is collaborating with Lucy. I grab at the guardrail to stop myself from tumbling backwards down the stairs.

Attentively, Ajit holds my arm and helps me regain balance. As soon as I am out of immediate risk he steps sideways to establish a polite distance from my body. He smiles sweetly and I can all but hear him thinking: *You are really clumsy.*

"Are you okay?" he asks, in a chirpy way which jars utterly with the distraught state of my body and brain.

"I'm fine, thanks for catching me."

"Sorry, I did not see you."

"*I* should say sorry, I was staring at the floor," I say and continue on my way.

In the lab, Lucy sits at a bench with a Bunsen burner, trying to pick single colonies from an agar plate.

"Hi Ka, how are you? Or more importantly, how was he?"

"I have a slight headache. And him… not impressive, boring."

"I told you that Hungarian pizza bakers are not good in bed," she says with an ironic tone in her voice.

"No, you didn't."

"I guess you had an inspiring breakfast conversation with him?"

"Yes. Riveting. Brilliant bit of action from you this morning by the way, cheers."

"Haha, how did he react?"

"Rather slowly. But he fucked off eventually."

While trying to get my moist hands into purple lab gloves I say to Lucy, "I literally bumped into Ajit just before."

"Ah, is he there?"

"Of course. Isn't he always in on weekends? Not at the moment though – just went home for lunch."

"That's good, I need to talk to him," she says.

"Are you not worried?" I ask her, while she places eight large vessels with broth on a tray.

She looks at me with question marks in her eyes. "About the content of your thesis?… I mean, Ajit is doing all the novel stuff. Your thesis will be very dull, no?"

"Yeah, it won't be too inspiring. Mark said during my second year exam it will be fine, though. But, yes, I'm worried about it."

"Is it really possible to get a PhD when you haven't done anything new?"

"Apparently so. I've basically just purified the same protein for three years."

"It was like a technician position, wasn't it?"

"Yip, it's been donkey work."

Her tone makes it plain I just asked her an extremely silly question, but I have abused my body and brain, and charm is not on my agenda today.

Lucy continues with a smile, "And right now I am purifying… a bit more! Still, soon I will be in Senegal, enjoying the sunshine."

Why is she so kind-spirited? I wish I could be so accepting and nice. I hate the world today.

I place my 50 mL falcon tubes in a rusty centrifuge, press the lid down with my elbow and set the timer to five minutes. "You think there might be a paper coming out?" I ask after sitting down on the bench opposite Lucy.

"Well, Ajit thinks so, but then again we all know he is a pathological optimist."

Lucy is right. Before she started to write up, Ajit visited our lab a few times per week to discuss research with her. He always informs Lucy that their work will soon be published in one of the best peer-reviewed journals; *Nature, Science* or *Angewandte Chemie*. "We are really not far from it, Lucy! We will soon have a very good paper together…"

"Sure," Lucy always replies in a weary tone that very clearly states; "Ajit, I love your optimism, you are sweet, but I'm just patronising you. Here, in the McLean lab, losing faith is a badge of honour. Our paper will never happen. And by the way, Santa doesn't exist either."

I tell Lucy, "Yes, Ajit is still dreaming… now that you're not there during the week, he sometimes tells me about your glorious paper."

"I told him recently there's a fair chance that even if we got enough results to draft a paper, it might well land on Mark's desk, where it can lie beside all the other ignored papers, to all intents and purposes resting peacefully like Sleeping Beauty for a hundred years…"

"Poor Ajit."

I take the tubes out of the centrifuge; I only need to isolate DNA from the bugs I just spun down. It's a very straightforward task yet I start to feel allergic to it – I have done it all too often by now. We chat about the night before and the Hungarian pizza baker who was in my bed this morning – it all seems sort of dreamlike now. We are laughing; I shake from the lack of sleep and the alcohol still making its way through my bloodstream. After a few minutes, and a couple of practical steps down the line, I realise that I added the solutions to my DNA extraction in the wrong order.

"Noooo! I can't believe it! Any moron can do this, but I manage to cock it up! Over and over again! Yet again I have to start from where I was two days ago."

Lucy clearly feels sorry for me, but she doesn't say anything. I sit down and put my head between my hands. "I can't be bothered starting again today, it can wait till tomorrow."

The tragic truth is: Today or tomorrow, it doesn't matter a shirt button. I do not even know if this whole experiment made sense in the first place. At the moment I have dozens of half-started projects, all given to me by Mark. With most of them I am not even sure anymore what exactly I already did. There are insane amounts of papers and books to read on my desk, which Mark keeps adding to. I decide to work on one project one day and on another the next. I cycle to the hospital to conduct experiments; sometimes they are experiments which I believed made sense when I woke up in the morning but then I enter the Royal Infirmary Lab and decide it doesn't make sense to do them after all. Usually I pop my head in to say hello to Sharon and Leonie, to find out about her latest Rugby injuries and hospital goings-on. Then I head back to the School of Chemistry without having done any actual work.

My mind is constantly spinning with thoughts, research ideas; trying to find out what is sort of promising and worth working on, what could potentially save my PhD… and while I am trying to sift through this mess for meaning, Mark pressures me to generate results for even more projects. It's a spiralling mess, but Mark will broker no discussion.

Most days I can rightly conclude that I might as well have slept all morning and spent the afternoon getting pie-eyed in the park with the winos. At least then I wouldn't have gotten stressed, and Mark wouldn't have gotten irked at my inability to distil his torrent of gobbledigook ideas.

"I'll come with you," Lucy says.

We pack our stuff, smoke a cigarette in front of the building and cycle back home.

"Again, why did you advise me to take the pizza baker home?" I shout to her, cycling behind me.

"I did not advise you to bring him home."

"You did! You told me it would be good to have a one-night stand!"

"It would. But maybe not with him!"

"Why is it a good idea anyway? This one-night stand thing?"

"Fun? To stop dreaming about this Alex? There are billions of reasons."

"I am not dreaming about Alex."

"Oh, you are!"

"Whatever…"

We park our bikes at the entrance of the flat and lock them onto the Victorian railings. My legs feel heavy when we climb the two floors up. As soon as I unlock the front door, my nose scents something bad. "Bah, it smells like him in here," I say.

Lucy laughs.

"Don't you smell him?"

"No Ka, I don't."

I remove all the bed sheets and open the windows but the smell doesn't seem to leave my nose.

Chapter 37

At 11:30 a.m. I am sitting behind my desk in the living room. I drink from the coffee standing in front of me, and it feels like I am sleeping with my eyes wide open. I am exhausted even though I slept deeply and lengthily. We had been out for dinner and a pub crawl, for Lucy's valedictory evening. Logan, Linn, Felix, Greg and William had joined us as we shuffled from one run-down hovel to the next. It had been fun, and intense, but we had got home by 1:00 a.m. and went to sleep quickly. Still, getting out of bed feels harder than on those many occasions when we drank well into the small hours. I feel physically tired and spiritually wearied. In recent weeks I have gotten increasingly nervous at the horrifying thought of finishing my PhD empty-handed, or even worse... not finishing at all. I wake in the middle of the night with my mind racing, like a frightened rabbit in headlights. I think about all the projects, and what a failure I am, and what on Earth am I going to do next, and if there is the slightest chance I can save the situation? ... I say I "think" but thinking is too active; this feels static and passive, like being devoured – by anxiety.

I open my laptop to check my email. One from Mark.

Karin,

I don't want you to spend so much time at the hospital. You should work in the chemistry department instead!!!

Best,

Mark

© Springer International Publishing AG 2017
K. Bodewits, *You Must Be Very Intelligent*,
DOI 10.1007/978-3-319-59321-0_37

I only think for a couple of seconds and then type my reply.

Dear Mark,

No worries about me working in the hospital. I am at home, not working at all.

Best, Karin

As I press the send bottom, I feel confident. Flashing through me is an unfamiliar feeling; pride in standing up for myself. It is the same as I felt when I dared to object to just one of Mark's silly research proposals. *A job well done!* I need to ignore that Logan or Linn or whoever Mark catches in the lab first might be the designated target of his random wrath. They won't be thrilled, that's for sure, but what can we do? It's just plain wrong to cow-tow to Mark; once in a while somebody inevitably takes a futile stand. Today it's me. I sigh, lift my head and smile.

Lucy notices and asks, "What did you just do?"

She is folding clothes to put in her suitcase. I let her read the emails. "You got your last will and testament ready?" she asks.

"I don't have any money or goods. But I can write down some old-fashioned ideals for posterity…"

"I'll take your chopping board instead."

"It's nice, isn't it?"

I feel liberated standing up for myself, even though it is pointless, counter-productive and likely to cause collateral I will feel guilty about. Glory is fleeting; troubles are long-lasting…

I get myself ready and cycle to campus. From the corridor I listen for Mark's voice in Lab 262. Sometimes I wonder if other supervisors in the department are aware of the dark, explosive tone which Mark deploys on his inmates when no outsider is around. Maybe there are people in the Department who feel sorry for us, or at least pity us. During the first 18 months of my PhD, Patricia Crick, who runs her own group on the ground floor, would occasionally pop her head in to see how we were getting on; always when Mark was at a conference or on holidays. But I haven't seen her for months. I was told that Mark and Patricia had collaborated on a project until Mark grabbed aggressively at Quinn's arm during one of their project meetings. Since then she never visited again. I don't hear Mark, so I open the door.

Both Logan and Linn look up at me as I walk in. "Seriously Ka, what did you write to Mark?" Logan asks.

"Why?"

Linn clarifies, "He came in before and he wasn't pleased. He ended up ranting for at least half an hour about the maintenance of the FPLCs and the mess in the cold room, but it was clear that it was about something you did. It was all quite unfair."

I tell them about the emails this morning and their faces yield to resignation.

"Can you maybe think about us the next time you want to do something like that?" Logan asks.

I know they are right. I feel bad about making Mark shout at them. I knew when I clicked on "send" that Mark would storm into the lab to vent. I knew it, and still I did it. *Why on Earth did I not phone to warn them?*

"I am sorry. Next time I will warn you."

"That would be appreciated," Logan says.

"If there is a next time…" Linn adds.

"Is he so angry?" I ask.

"Hell yes, he's furious!" Linn says.

I sit down behind my computer and start to search the Internet, though I don't know what I'm searching for. A few minutes later I am sitting at the Bunsen burner, picking some colonies from an agar plate that I spread yesterday, when I hear Mark enter. Instantly I consider hiding behind the fridge. But he would find me and I would look silly.

Mark pauses next to me ticking with his key chain on the bench and releasing a very audible sigh. I am visibly nervous. I know that this man – who I know to be nothing more than a bitter individual – is going to lacerate my psyche at any moment.

For what feels like a long time, in reality probably just a few seconds, he looks at me without saying anything, as if he doesn't know what to say. He could ask me about the email – which would be directly going to the heart of the matter – but he doesn't. Eventually he barks, "I want your results on the ACP project on my desk this afternoon!" *ACP?*

"I don't have any results on that project. I have been working on KdtA, as we discussed last week."

"ACP I asked you for!" he barks again.

His eyes roll, he shakes his head in disbelief and ticks his keys even louder on the table. I feel my hands stopping to shake and the strange nerves in my belly disappearing. I open my mouth and want to shout back. I want to tell him that he is wrong. *Don't shout back. Don't shout back. You are above all that. And fuck, you need to get your PhD! You need to get your fucking PhD…* I feel my shoulders sinking and a deep feeling of sadness overtakes me. I look him straight in the eye and for the first time, in the presence of Mark, there is a tear running down my cheek.

He looks at me for a few seconds. Surprised. Confused. "Sorry," he says, then buggers off.

Linn walks towards me as the lab door falls close. "Arsehole," she says.

"He is," I say and walk to the office.

I fetch my coat and dig in my drawer to see if I have a forgotten pack of cigarettes somewhere, but there is nothing. I go outside in the hope of finding a fellow addict who could spare me a cig, but there aren't any. I head to the Simpson lab, beg one from Felix and walk back outside. Prof. Gilton is there now, puffing away.

"Karin, nice to see you!" he says, in his informal and happy way.

"Hi," I say, much more reserved.

"We need to arrange a date for your second year viva. *Soon.*"

Every PhD student at the School of Chemistry sits two exams before the final graduation. I knew that the second year viva is not much of an exam; it's a mere formality to check you are on track with your project. I was supposed to have mine last summer, but so far Mark has postponed it.

"I asked Mark a few months ago, but he said he is busy and it can wait."

"No, it can't! We need to make sure you finish in time. I will talk to Mark and arrange a date."

"That would be great," I say, hoping this exam might clarify what I am actually working on.

I come home early evening to find Lucy stretched out on the sofa smoking a cigarette, with a large suitcase next to her. I sit on the window sill, resting my feet on the rocking chair, and light my own cigarette.

"How did Mark react?" Lucy asks.

"SSsshh…" I hiss, indicating that I don't want to talk about it.

Lucy nods, understandingly.

"Are you not having a date with ice cream man tonight?" Lucy asks, referring to a guy I gave my phone number yesterday evening in the pub.

"I had him on the phone this afternoon. He sounded like a total schmuck. I told him I've got swine flu."

Lucy looks at me and laughs.

We sit next to each other without saying much. It is unusual for us not to talk when we are in the same room, but I guess we are both mentally preparing for her departure. To break the silence I walk to my laptop on the small desk, connect it to the boxes and call up John Denver's *Leaving on a Jet Plane*. We both laugh and start to sing along.

When the song ends Lucy stands up and gets into her coat. I follow her to the corridor to say goodbye.

"First Vlad and now you…" I say, feeling close to tears.

Of course, having Vlad leaving a couple of months ago was nowhere near as bad as Lucy leaving now, but still it had been a painful goodbye.

"Oh, come on now. You have no reason to be so unhappy. You have great people around you. Felix, Greg… if you need some crazy input, William, and if you get really bored you can always ring the Hungarian pizza baker or Alex," Lucy says.

"You know Alex is off-limits," I say, forcing a smile. "And the pizza baker was a been-there-done-that-never-again experience."

"I will come back at least twice. Once to submit and once for my PhD defence."

I throw my arms around her and give her one last hug before she walks down the stairs. "Have a good trip!"

"Take a stand, alright? Don't work for free and stay strong! Promise?"

"Promise!" I say wondering how on Earth I will stick to that, feeling as lost and bereft as I do now. "Think about me walking through the Meadows in the drizzling rain when you sit in the sunshine."

"Will do," she says and walks down the stairs.

I do not close the door of my flat until I hear the front door of the building fall close.

Less than two hours after Lucy leaves, the doorbell rings. I am not expecting anyone but I buzz them in and wait at the front door of my flat to see who climbs the stairs – Felix and Greg. "Hi," I say enthusiastically.

"Surprise!"

"Sure is."

"We were just round the corner and thought we'd call in and say hi," Felix says.

I suspect that there is about a zero percent chance that they just happened to be round the corner – what the hell would they be doing there, *together* – and they are here solely to cheer me up, which is touching. I feel grateful.

"You want something to drink?"

"Oh, we brought something," Greg says and opens a backpack full of different types of Belgian beers plus two bottles of wine, white and red.

"You don't mind that we just stopped by, do you?" Felix asks, already seated on one of the benches in the kitchen.

"Not at all!" I say, feeling rescued from depression by their not-at-all spontaneous visit.

"We got something for you!" Felix says, handing me an envelope.

"What is it?" I ask, curiously ripping it open; it contains three tickets saying *Hearts versus Rangers*. "Football?"

"Yep!" Greg says. "We know you're not into it, neither are we. But let's just go for the atmosphere..."

Felix adds, "For the sheer joy of feeling that you are part of something and you are breathing!"

"Sounds great!"

Chapter 38

"Mark is in my office, eager to start," Prof. Gilton says and I think I detect a note of sarcasm in his tone.

Though Prof. Gilton has, so far, always been ever so correct during our chats between the chemical waste containers, I find it impossible to imagine that he does not hate Mark. I have no evidence that he does but I cannot believe a decent person wouldn't. But maybe that is my problem, after all Mark was his PhD student, and Prof. Gilton seems to be his mentor…

"That's great news," I mutter and stand up from my desk to follow Prof. Gilton to his office, holding onto an A4 piece of paper with my notes.

I try to hide how distressed I feel. "It will be fine," Prof Gilton states, just before he opens the door of his office.

I am not sure if he means that he will not grill me in the way he did during my first year exam or if he is hinting that he won't let Mark cook me alive. The door opens and Mark has a smile plastered on his face. To me it is laced with horror-movie malice but maybe he wants to put on a good show for Prof. Gilton – I hope he does. I shoot him the most sympathetic face I can manage and sit on a chair on the other side of the table. I feel my heart pounding fast. I am overwhelmed by a desire to mentally shield myself from any humiliation which might follow. In the two and a half years I have been working for Mark I have had, at most, a handful of normal conversations with him. The rest has just been him bulldozing me with one of his monologues, shouting at me as the mood takes him or admonishing me with contempt.

Prof. Gilton sits on a chair at the side of the table, strategically placing himself between me and Mark. He quickly explains the aims and formalities

© Springer International Publishing AG 2017
K. Bodewits, *You Must Be* Very *Intelligent*,
DOI 10.1007/978-3-319-59321-0_38

of the second year viva and emphasises a few times that the final goal is to define a clear end-line to the PhD and to discuss the table of contents of the thesis.

"So, where are you standing now and what can you still finish?" Prof. Gilton asks calmly, looking at me.

I take a deep breath and look at my notes in order not to say anything unplanned. "So, I have many projects going on. I think three might yield good enough results to write a thesis about – assuming I am not given any more projects."

I explain which projects I have in mind and that I believe, with the available equipment and lack of research funds, the only project that realistically might still lead to a paper is the work I took over from Erico when he left. Mark is sitting with his arms crossed, head shaking, clearly disagreeing with every word I say. A few times he tries to interrupt, but Prof. Gilton always gestures to him to stop, saying, "Let her speak first." Prof. Gilton is writing, crafting a table of contents for my thesis. He asks a few clarifying questions about each project.

"What about the LptA…, and the LpxC project?! I even *bought* you the plasmid!" Mark asks aggressively.

It was true that Mark finally bought me the LpxC plasmid, but unfortunately the protein doesn't want to express in *E. coli*. I am replaying a conversation with Lucy in my mind: *"Take a stand, alright? Don't work for free and stay strong! Promise?" "Promise."* I take a deep breath and say: "The plasmid doesn't work. And it is all too much. I want to be out of the lab by the end of June, so I still have two months to write up."

"What?" Marks asks, as if this is the stupidest thing he has ever had the misfortune to hear.

"Shsst," Gilton hisses, giving Mark a warning sign with his finger.

He starts counting on his hands how many months this boils down to, looks back at the notes he made, and concludes: "With five months of lab work to go, the projects you have in mind sound plausible to me."

"She has more than five months to go!" Mark splutters.

I open my mouth to speak, but Prof. Gilton is speaking instead: "Her stipend runs out at the end of August?"

"Yes, but that doesn't mean…"

"Shsst," Gilton interrupts. "So five months lab work plus two months writing it is." *Thank you good man, thank you, thank you…*

"Are you joking?" Mark asks, looking at Prof. Gilton.

He has not properly started shouting yet but his eyes are bulging – a sure sign the detonator is set.

"No, I am not."

"This is ridiculous!" Mark is properly raising his voice now.

Prof. Gilton slaps hard on the table. "Enough! This is how we are going to do it! Period!!"

Mark opens his mouth to utter another protest, but as he looks at Prof. Gilton's face he decides to take his losses and nods resentfully.

Prof. Gilton prepares the paper work, writing up exactly what we have just agreed upon. Mark and I wait in charged silence. All three of us sign a copy.

Prof. Gilton speaks with the same calm voice he had at the beginning, "Thereby the content of your thesis is standing. You know now what you still need to do. However, as you know by now, research is not predictable. In case you run into issues, you can come to me and we can discuss it again."

He stands up and both Mark and I do the same. Before I walk out of his office, Prof. Gilton lets his hand rest shortly on my shoulder.

"Thanks," I mumble.

"You're welcome."

I instinctively walk to Felix's desk to tell him about the meeting, but then remember he flew back to Denmark for a funeral. I opt for Greg's office instead and cross campus to the School of Mathematics. I enter the large, silent ground floor office with desks and numerous computers. There are a few people working, completely focused on mathematical models predicting phenomena I don't understand. "Hi, petal!" Greg says, lifting his head from the rugby match he's watching on screen.

"Working hard, I see."

"As always. You're early for social time. That will be very confusing for the people here. To move this by an hour, oh, a very big deal…"

A few weeks previously I got into the habit of dropping by late afternoon, mainly to score an espresso from their first-class machine, but also to chat with Greg. The silence of all the number nerds in the office felt terribly awkward, and I was curious to see what kind of people they were, so I started calling out "twenty minutes social time!" on my arrival. The first day I explained what the idea behind it was – that we would have a break together – and on the second day all inmates obediently left their desks. They were far from enthusiastic; in fact they were rather lackadaisical about "social time," shuffling indifferently towards the coffee corner like bundles of clothes being swept along by a giant invisible brush. However, by now I know all their names, except for the one postdoc everyone refers to as "Number Three," on account of him having been the third candidate interviewed for the open position. It didn't seem to bother Number Three that he was called Number Three, so it stuck.

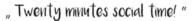

"I'm not here for social time yet. I just had my second year viva!"

"How did it go?"

"It was weird, but the outcome seems really good!"

I tell Greg about the disagreement between Prof. Gilton and Mark, the shouting and the paper we signed. "Oh Jees, Mark will be pleased," Greg states.

"I know. I guess this can go two ways; the good way, meaning Mark doesn't talk to me anymore. Or the bad way, meaning he unleashes hell on me every day."

"It seems Gilton is on your side. Is he the alcoholic Prof.?"

"How do you know he's alcoholic?"

"It's gossip all across campus…"

We chat a bit longer and I walk back to the chemistry department. I copy the note with all the points three times and stick one of them on the side of my computer. *That will remind Mark every day when my years of servitude will be over… lest he suffer from Alzheimer's…*

I stare out of the window and spend at least five minutes reminding myself that getting my degree will be doable and the submission of my thesis is only a few months away. Surely I can manage it; going through this a bit longer to get the hallowed title – even if it does seem somewhat tarnished now.

Right now I am emotionally exhausted and just want to beam myself to my parents' sofa; there I would passively watch a farmer ploughing the land behind our house in preparation for sowing seeds… The village seems far away in another world, another life, it seems idyllic too, but I left that behind, for *this?!…*

Chapter 39

Spring has manifested in a yellow glow, from the numerous gorse plants in and around the city. The summer semester has started, and campus is filled with young and lively undergrads. Babette has left the lab and is now writing up. With another inmate down, it is only Linn, Logan, Barry and me left in Lab 262. Linn and Logan are dating each other, which is kind of cute. It is also something of a relief after the long pussyfooting love bird phase of will-they-won't-they-oh-just-get-on-with-it-and-fuck-each-other-senseless-would-you. We all came to think of it that way but I noticed the pussyfooting phase particularly riled Babette. Barry is not in the lab much, unless there is a student to supervise. That sits well with me and, I suspect, some others.

Mark hasn't uttered a single word to me since the exam. It is uncomfortable to be so completely and utterly ignored by one's boss but it is joyous compared to the alternative; being shouted at and perhaps fired at the eleventh hour. The exam has motivated me. Despite finding myself staring unproductively out of the window several times a day, questioning what I am doing with my life, I am working hard to meet the agreed targets.

Best of all, over the last three days it appears that the project I took over from Erico really is bearing fruit. I have Peter, an overly enthusiastic and – thankfully – very competent bachelor student, spending hours and hours on the bench with me; we are both convinced we're onto something. Maybe we have indeed found an old, almost forgotten antibiotic and we might just lay the foundations for its novel use; in treating cystic fibrosis patients with fatal infections. Of course, it would need many more years of research,

© Springer International Publishing AG 2017
K. Bodewits, *You Must Be* Very *Intelligent*,
DOI 10.1007/978-3-319-59321-0_39

clinical trials and God knows what, followed by many bureaucratic steps that could stall the project forever, and certainly well before mankind might actually benefit from it. But even so, it is promising. The results are still in their infancy and more work needs to be done, but it feels good to have hit on something just before the finishing line. The end of my PhD is nearing and it might yield something of genuine interest after all – I am happily incredulous.

Giving a final presentation of my work, at a spectacular location in the Highlands, in front of the bio- and organic chemistry section, is a box that still needs to be ticked – a compulsory part of the PhD curriculum. It's potentially a good opportunity to test the waters with ground-breaking research. But unfortunately I haven't got anything to present yet. Alas, I cannot move the dates; I am compelled to drive north bereft of interesting results.

Logan, myself, all other second and third year PhD students, and three members of staff gather in front of the department to drive up for the three-day stay at Loch Tay, about 85 miles north-west of Edinburgh. There is just a seminar room, a canteen with view over the loch, a few bedrooms with wooden bunk beds and a cheap makeshift bar in the basement serving beer and wine. It is Spartan bordering on tawdry, but the gorgeous, dramatic scenery more than compensates. It should be a highlight of the PhD programme. I loved it last year when I had just been in the audience, but this year I am nervous about the paucity of my presentation.

"Jees, Ka, you aged by ten years overnight. What happened? You went to the pub?" Logan asks when I walk towards him.

I have a heavy backpack on my shoulders and am holding a takeaway coffee from the new coffee stall parked on campus.

"No. I spent last evening assaying proteins, was home at midnight and got up at 4:00 a.m. to get here in time and give Peter instructions for the next few days."

"Hardcore motivation, haven't seen you like this for *ages*."

"I know…"

"Who're you driving up with?" Logan asks smiling at me, standing next to his small car.

"You?" I say, a desperate look in my eyes indicating there are not really many alternatives.

He knows damn well that I had not replied to a single one of the emails going around concerning practicalities like travel to Firbush. I just hoped he had kept a seat in his car for me, which he had. He is a good man. He understands my motivation has limitations.

Together with William and Jonas, the Swedish guy, I get into Logan's old Fiat Panda. We all need to keep our backpacks on our lap as Logan's has filled the little trunk. I move my seat as far as possible to the front so the guys in the back have a tiny bit of space as well.

"Yo Billy, how are you doing?" William asks Jonas as we are driving off. *Oh Jees, good start.*

Thankfully, I see Jonas smiling in the back mirror, otherwise this could be a long-long trip.

"Still think I am named after an IKEA wardrobe?"

"Yep. Looking at your face, I even think you walked into one this week." *What the hell!*

"I hope you are referring to my black eye?" *Please say yes now.*

"Yip."

"That was a baseball."

"Why are you actually coming to Firbush?" I ask William to change topics.

He is neither staff nor a PhD student. "My boss, the dude you refer to as Homer Simpson, considers the Firbush concept a burden rather than a jaunt. Hence he sends me instead."

"The bottom minion in the pecking order gets dumped on..." Jonas says.

"Ho, ho, ho, Billy, mince your words!" William says, laughing.

When we are well into the greenery and the roads are narrowing with tight, scary bends, Logan asks: "Did you hear Mark is getting a baby?"

"No! He scrapped me from the to-talk-to list"

"Apparently it's due in three months."

"Maybe it will change him in a positive way."

"Hopefully."

Nobody is really hopeful.

On arrival we are warmly welcomed by our seminar hosts. They outline the programme and tell us that between presentations, they offer sailing classes, hikes, windsurfing and so on. Last year, I had gone on a memorable hike with Lucy and a few girls from the Johnson group, into the hills around Loch Tay. But this year the weather is dismal and I guess I will mainly bunker down in the dorm room that I am to share with three other female students.

I recognise the faces of my room-mates, who all seem nice, but I don't really know any of them. However, one of them is the ex-girlfriend of Thomas – I recognise her from the pictures in his flat – who is now in Canada. Does this girl, with whom I am sharing a bunk bed, know that he loved her so much that he turned his room into a shrine of worship for her?… Maybe he sacrificed every single picture of her in ritual fashion before departing for Canada? But maybe he also took her pictures – and underwear – across the ocean with him. I would love to know if she knows about this, but it seems a strange subject to raise with someone you barely know. *"Hi, did you know your ex treasures your used undies?"*

We have lunch together and then the first session of presentations commences, which lasts until the early evening. Some of the presentations are just good, others are very good. I am increasingly nervous. It will be me on stage tomorrow and I know, for certain, my presentation sucks.

After dinner, many head for drinks in the basement of the building but I hurry to my laptop in the dorm instead. *Is there any way to pump up my presentation and add the latest results?*

I scroll through piles of photos of antibiotic assays I made in the past few days, but as the critical measurements of the plates are only in my lab book in the hospital, it doesn't make sense to add anything. The protein purification graphs and assays are all stored in computers in Lab 262, only some of which are accessible via a floppy disk drive. It had been too hectic to bring it along, and even if I had all that… they would doubt my results before I repeated the experiment twice. I close my laptop and lie in bed with my eyes wide open. I hear the other girls entering and remain silent. I stare at the roof for a long time before falling into anxious sleep.

The room breaks into feeble applause and everyone seems to be wobbling on their seats with unease. The blond, Spanish PhD student, Elena, who normally always smiles, walks back to her place in the audience. She looks utterly defeated and humiliated. Her face is red and it's possible she's about to cry. If I were her I would probably cry too. Maybe not here and now, but later when I returned home I would sit at my window sill and weep. She gave a good and enthusiastic presentation, but she was fried alive in the Q&A afterwards. It ended up in a public argument. I have never seen anything like it. One of the members of staff, who has just started his group in the chemistry department, bit into her work like a shark that wouldn't let go. He made it plain that what she was presenting was simply all wrong, that she had used the wrong techniques to prove what she wanted to prove and that her interpretation of the results was entirely inaccurate. I know too little about the topic to judge who is right. The other two young group leaders and William nodded along with his comments indicating they were onside. They did not believe in the work Elena was presenting either. *Is it plausible? That Prof. Johnson, one of the two gilded robes at the school of chemistry, would let a final year PhD student from his group present research that is fundamentally wrong?…*

I have heard that with group sizes increasing to well above twenty people it gets very hard for supervisors to maintain meaningful personal contact with every single student. Plus, the big shots spend a lot of time hanging around at conferences; some are hardly at university during semester breaks. If you are lucky you have a postdoctoral researcher in the group working on the same or a very similar project, and they guide you through your PhD. But if you don't have that, you can become like a ghostly hermit doodling in a secluded bubble. *Poor, poor girl. At least I don't have results to grill me on. I still have nothing, nada, not a sausage…*

I feel my heart starting to beat faster and cramps shooting through my belly. Only one more presentation to go and then it will be me on stage; the final presentation of the morning.

Anton is on stage now, a PhD student working in the Simpson lab. I have spent some evenings in a bar with him and of course I see him often enough while fetching a coffee. He is a pleasant, sociable chap who can down a pint of beer within three seconds, and he is physically fit and apparently an excellent researcher. He managed to publish a paper in *the* "tabloid" journal of science, *Nature*, during the second year of his PhD. I've never seen him present before, but Felix told me he is good. Still something desperate and nasty inside me hopes he will flop miserably; so that the contrast between us is not too brutal. Anton opens his presentation on the laptop connected to the projector next to the podium, comes a few steps towards us and starts to talk. After only a few sentences I am blown away by his speech.

He talks confidently and engagingly. His presentation is designed with precision, and he manages to explain everything so well that even me, coming from a totally different field, understands every single word of it without getting the feeling that this is a dumbed-down presentation for idiots. He presses the button to open his next PowerPoint slide which is a movie, no less. Colourful complex molecules walk along a string, calling to mind a sloth on a tree branch. It is perfect and impressive. I quickly scan the faces of the three staff members in the room; they all look suitably pleased. *How did he do all that? And precisely why on Earth is it miserable me talking after amazing Anton?*

He keeps talking with passion for another twenty minutes after which the audience erupts into applause. I feel my heart pounding in my chest as if it wants to burst out of my body. I could only think about how this all happened. *Why had I accepted this hateful PhD position and why did I not leave the moment I realised it wasn't leading to anything? How could I have gotten myself into such a humiliating situation? What are all these people going to think of me?*

Anton is still answering a few questions from the audience. My head is buzzing, my nerves are fraying and my brain can no longer process Anton's words. My hands are trembling and I feel sweat coming out of every pore of my body. He rounds off the Q&A, receives his second applause and makes his way back to his seat.

"It's you now," says Logan, sitting on the chair next to me.

I stand up slowly. My legs feel wobbly. They move forward uncertainly and seemingly of their own volition. I climb the two steps to the podium and am in a state of shock. As I walk to the laptop, I am conscious of an eerie silence filling the room. Everyone is waiting for me to open my PowerPoint slides. My vision is blurred and my fingers are so sweaty they stick to the mouse pad. A typical Windows desktop with a little white arrow jerkily jumping around the screen, unable to reach its goal, is projected on the large white wall. I try to dry my fingers on my skirt. When I reach out for the mouse a second time Logan is standing next to me and opens my presentation for me.

"Thanks," I hear myself whispering hoarsely.

Never have I felt so nervous about a presentation before. I used to quite like talking in front of an audience, but today I am overwhelmed by the silence and the number of people watching me. I am no longer secure and confident academically, and apparently not as a person either. If this would have all been a week later with more results – maybe then – but not now.

I look back at the audience, catch a few eyes and force myself to smile. I am searching for the starting sentence in my head, which I had prepared for this presentation, but I don't find the words. Desperately, I look at the first PowerPoint slide showing the title of my talk and decide to read that out instead. I hear my own shaking voice reverberate in my ears, sounding weird and gurgley as if I am hearing myself underwater. I keep on talking; the words seem to be flowing out of me without my control as if a radio is playing in the background. I flick through the PowerPoint slides one by one explaining the things I have done in the last three years. It appears to be beyond me to even try to make the work understandable or inspiring for the audience. Within a few minutes I am conscious enough to observe that everyone is drifting off in much the same way that I drift off during boring presentations. I charge through the presentation apologetically, genuinely sorry for detaining everyone, especially myself. I get to the finishing line in half the allotted time and everyone seems suitably relieved when I say "thank you".

The applause is, of course, embarrassingly weak. One of the young lecturers stands up and faces the audience, "Does anyone have questions for Karin?"

"Why are you here?" might have fitted the bill, but the room just fills with a silence which conveys more clearly than words can, "let's just get this over, of course I wasn't listening, nobody was…"

Normally, if no one has a question and the professors have been dozing during the presentation, one of the senior scientists steps in with a face-saver such as "Did you try this reaction in a different solvent as well?" or "What are the potential applications for this work?" But not this time. The audience is resolutely silent, lest I forget just how amazingly dull I was.

I manage to regard the expressions on people's faces but I'm not sure if they simply hadn't followed a word I said or if they just feel pity for the dull Dutch girl. As the silence intensifies and I start to feel even more undignified and exposed, like a fat stripper at a posh Gentleman's Club, the short guy who just grilled my Spanish peer says: "So, it's cystic fibrosis you're working on, isn't it?"

"Yes," I say, nodding in his direction.

"McLean group, eh?"

"Yes." *Wow, technically speaking, my talk even inspired two questions!*

He nods, suggesting that one can expect no better from a McLean inmate.

"Very interesting," he says, valiantly trying to hide his complete lack of interest in my research. "If no one has any questions then thank you very much."

Slowly I walk off the stage, during which everyone is informed about the lunch and what time we will start again in the afternoon after our outdoor activities. I guess it was empty stomachs as much as anything which saved me and my dirt-poor presentation from tougher scrutiny. Right now there's nothing I want more than to go home with a bottle of wine and down it on the window sill. But I am not even close to home… Everyone heads to the canteen except me who makes for the bar in the basement. The door is open, but there is no one inside. A middle-aged lady with freshly coloured hair comes out of the little staff room at the side.

"Can I help you with anything?"

"I wondered if I could have a drink."

"The bar isn't open yet."

"Could I secretly get one?"

The woman looks me in the eyes, presses her lips together, and nods. She unlocks the door and hands me a beer from the crate behind the bar.

"Thank you. I will drink it outside."

She doesn't say anything, just nods conspiratorially.

I climb the stairs out of the basement to be back on ground level and head to the Loch. I sit down on a large stone close to the water, far enough from the conference centre not to be seen, and take a monstrous gulp of beer. The fog hanging in the valley over the water slowly starts to burn off for the first time since our arrival. Normally, I would be happy to see the gloom of claustrophobic water drops vanish, in the magical way of the elements. But today the bright world revealed is not something I want be part of. Firbush is a lovely place but this year it just bears me down. I feel small and, as ever, I wonder how it came to this?

Chapter 40

It is just after midday when I enter the lab. I had been writing my thesis this morning, when I realised I was mainly staring at a blank screen. So I had put on my backpack and cycled to university through mid-summer Edinburgh. There are still a few small experiments to finish for my thesis, to round off the story. There is also data that needs to be processed and I don't have the right software at home. Alas, I still have to come in for a few days.

"Karin, can I have a word with you?" says Barry with a trembling voice. *He would be bullied so much by teenagers if he was a high school teacher – or a high-school student.*

"Sure," I say, much more enthusiastically than I feel.

It's an ominous phrase for me. I doubt any conversation that begins with "can I have a word with you" ever ends well. I am completely mystified as to why Barry would open a conversation with this condescending remark.

He is the postdoc and I am the PhD student, but that does not make him my superior. We had recently chatted about the mess in the lab, and with drooping shoulders he had politely requested that I participate more in cleaning up. True, I had gotten lax, but I tried to change after we talked. These days I am hardly in the lab. *Surely he isn't going to get all sergeant major on me about mess that isn't mine?*

"Let's go outside," says he, now sounding uppity.

"If you like."

Is he actually taking me outside to tell me off? Beyond outrageous. He's flipped...

© Springer International Publishing AG 2017
K. Bodewits, *You Must Be Very Intelligent*,
DOI 10.1007/978-3-319-59321-0_40

We walk to the few wooden benches behind KB House and I sit down. Barry does not. He remains standing. A charged silence ensues. He is extremely nervous, and it seems to be contagious; now I am nervous as well. *What the hell is this about?*

"You want a smoke?" he asks, holding a pack of cigarettes in front of me, with a hand that would suggest he has severe Parkinson.

I take one and light it with one of Barry's matches. He lights one too and inhales deeply. This seems to fill him with confidence. He looks at me, angry now, which is really kind of silly but, nonetheless, he looks concentrated like a tiger focusing on its prey. If it weren't so public I think such a personality change might scare me. I still have no idea what is going on.

He takes another extraordinarily deep drag from his cigarette, which he has now half-finished with two magnificent puffs. Without a word of explanation it is already plain as day this will get nasty and threatening very soon. He doesn't hold a weapon, but it feels like he does. In some way, it seems kind of absurd, even comical; Barry playing the role of scary mean guy, which he is just not cut out for. I think of him replacing Anthony Hopkins as Hannibal Lecter and that makes me smile. The smile disturbs Barry. He wants me intimidated, that much is clear.

"Uh… about this article you're writing," he says, opening his mouth a bit too far, moving his lower jaw from left to right, as if he isn't yet sure what words to use next.

"It's already finished. What about it?"

I speak guardedly, but I am eager to know where this can possibly be going. The article concerns the project I took over from Erico and it has nothing to do with Barry. He did help Peter when I had been in Firbush, plus on the one morning I had been too hungover to come in. And we did ask him for advice on the protein assays when we got stuck, as he is supposedly "an expert" on that, but that was it.

"I want to be first author on that paper," he says.

His voice is astonishingly firm, yet the words are ludicrous. I laugh at him, uncomfortably at first but then rather hysterically.

"There's not much to laugh about, Karin," Barry says, peering at me.

"Well it's ridiculous, isn't it? Why would I put you on as first author, or any author?"

"If you don't, I will get you into *big* trouble with Mark. And you know I can do that. I'm on good terms with Mark and you aren't. Guess who he is going to believe?"

I look at him, flabbergasted.

"So you're going to make something up about me?"

"We both know I can do that." *Oh hell yes, you could. And you even have such a passive, sad beaten dog reputation that people might well believe you. But... even so Barry, go fuck a donkey...*

I stare in front of me in silence, watching Fatso, the fat cat, under a bush sleeping next to something that I suppose had been a cute rabbit until a few hours ago. Barry is clearly waiting for me to respond, but I don't. I stand up and walk away from the silence, across campus in the direction of the School of Mathematics, thinking I might explode any moment. My tongue is moving, but my lips stay closed. I repeat all the insults I remember hearing at the football match: you frustrated fuck trumpet, shit gibbon, bloviating flesh bag, and fuck the conflict management course I just wasted my time in, bloody useless theory…

"Hi, flower," Greg says, turning towards me.

He has his arm in a bandage, his sausage fingers sticking out. Greg, like Sharon, spends much of his life suffering from severe rugby injuries. If it wasn't that Sharon already had a partner, they could have been a good match; opting for assisted living in a house with a disabled toilet as soon as they move in together.

"Hi." I can't quite speak yet.

"What's the deal, petal? Twenty minutes social time?… Oh God, you look furious."

I tell him what just happened. "Shall I come to your lab and beat up doctor Barry?"

"With one arm?"

"For Barry? A pinkie ought to suffice…"

Greg is at least a head higher than Barry, and weighs about sixty pounds more. He does rugby, Barry does depression. We both smile, though I seriously like the idea.

"Why would he do that?" Greg asks after a few seconds.

"I guess he's getting desperate, more than two years into his postdoc and no significant results."

"That kind of sucks, but it's no excuse."

"I don't know what to do."

"Why don't I get us an espresso?"

A moment later, Greg is back with two tiny cups of espresso. "Why don't you go to this Prof… the Gilton guy… ask him how to deal with the situation? From what you've said, he seems reasonable."

"I suppose that's a better idea than punching Barry in the face, but less gratifying."

"You can still do the punching afterwards."

When we finish the coffees I head towards the chemistry building, still feeling stressed, bewildered and violent.

As I approach Prof. Gilton's office I think briefly about what it had been like when Barry started in Lab 262. He had never been keen on an academic career. As far as I remember, he signed up for the postdoc because he couldn't find a position in industry. I suspect the postdoc was supposed to be a sort of parking place until he finds a job he actually wants. But now he owns a flat here, the economy is in tatters and the pickings are poor for scientists in Edinburgh. Maybe he needs a paper for his academic career now? He has certainly been very distressed of late. But that is as inevitable as night following day when you are dating Babette. *Babette! No single punch in the face would be hard enough to rival the pain wrought by Babette. Poor guy... No, he is an arsehole!*

"Come in," Gilton shouts in his husky voice.

I enter the small, cosy office.

"Sit down," he says, stuffing an ashtray into a desk drawer.

I take one of the two free seats on the other side of the table. "What's behind those eyes, young lady?" he asks benignly.

I want to start my sentence with "the frustrated fuck trumpet postdoc" but I am determined to sound non-violent. I convey what happened as if I am not seething with murderous rage.

Gilton listens attentively and when I am finished, he takes his glasses off his nose and lays them on the table next to him. "First of all, these kinds of things happen much more often than you think. And sometimes there is very little you can do about it."

"But... it is unfair."

"Barry probably feels his own output isn't enough. His contract will soon run out and then the decisive moment will come: he either makes it in academia or he doesn't."

"But that is not my problem, right?"

"No it isn't. Don't care about him. But care about yourself, Karin. You only have a couple more weeks before your own contract runs out. Your focus should be on getting out of here. That's much more important than this paper."

"But I don't want to give Barry a paper just because he puts a knife at my throat."

"Then don't. Just let the paper be. Or are you planning to pursue a career in academia?"

"No! I don't have a chance anymore. But it's months of work. Plus, it was Erico's project. What about him?"

"Erico is digging for oil nowadays. I doubt he cares."

"Hm, true," I reflect. "It's already on Mark's desk, has been for a few days."

"Well, don't bring it up. I strongly doubt he will."

Gilton looks at me piteously. "How's the thesis going?"

"Got most of it finished actually."

"You got someone proofreading it?"

"James reads it and gives feedback."

"Mark?"

"Not so far."

Gilton sighs, irritated.

"It's always the same with him," he says wearily. "You've got all the bullet points done that we agreed on during your second year viva?"

"Yes."

"Very good. If you run into trouble just come to me, Karin."

He stands up to indicate that we are finished.

I leave his office, and walk to the lab. Barry is standing at the UV-vis talking to one of his new project students. I look in his direction, but he's avoiding eye contact.

"Frustrated fuck trumpet," I mumble.

Chapter 41

I take my whole drawer out of the fridge in which I have stored the largest part of my samples over the past three years, and place it on the bench. One by one I go through all the tubes. On many of the older samples the text on the lids has faded and I have no idea what is inside. Others I didn't document properly and, despite the text being crystal clear, it could be pretty much anything in there. But, with most I know exactly what is in the tube and if I dig deep inside myself I can recall the anxiety-arousing memories of what it had taken to create these samples – memories I have been managing to suppress rather well lately. Even though I only stayed home for two weeks in a row to finish writing, the experimental part of my PhD seems like years ago.

Carefully I place a few stocks of the plasmids I created in the common Lab 262 plasmid box. I sigh deeply and let all other samples drop in the large yellow bio-waste bin next to the bench. I follow the same procedure with all the stuff I kept in the −80 degrees freezer; only three glycose stocks with modified bugs I place in the common box; the rest I unceremoniously dump in the bin as well. I am looking down at the 100s of samples, feeling a mixture of pain and relief to see all the hours I invested being thrown away – it's closure on a sad waste of time.

I already tidied up my bench in the hospital a few weeks ago, when I said goodbye to Sharon and Leonie. James had officially retired and despite him still coming to the office every day, his lab had closed. Leonie started a new postdoc in England and Sharon had found a position as a technician at the Roslin Institute, where Dolly the sheep was famously created. Brian moved to Cork.

© Springer International Publishing AG 2017
K. Bodewits, *You Must Be* Very *Intelligent*,
DOI 10.1007/978-3-319-59321-0_41

I tidy the cold room and when I think I am finished, I walk to the office. My desk is empty; a few weeks ago I took home all the papers to write up my thesis.

"Ready?" Logan asks.

"Still need to submit," I reply, happy in a strange way that it is not yet time to say goodbye.

I check the clock to see if the College Office has opened; another ten minutes to go. The time hangs heavy so I decide to take a few filled-out forms and the two heavy copies of my thesis, along with a USB stick (containing the same file electronically) to wait at the Office entrance door. I enter as soon as it is unlocked. Two friendly ladies greet me and I place everything on the counter. It takes less than five minutes and I am standing outside again, much lighter in every sense.

I walk back to the lab. I had thought the chances of still seeing Mark before my departure were slim. Linn told me they heard from Barry that Mark became a father last week and they had not seen him since. But now he is there, standing with sad Barry at the UV-vis.

"Logan?" he shouts into the office, in an annoyed tone.

"Yes."

"What happened with the software on this computer?"

"Oh, I don't know," Logan replies while walking towards Mark and Barry, who both completely ignore my presence.

A few seconds later Logan is at the machine. "Congratulations, Mark, with your son!" Logan says.

He extends his hand to shake Mark's, who tries to keep Logan's hand at a distance and looks irritated by the gesture and topic.

"Are we now talking about babies or about this software?" he barks aggressively. *Sociopath.*

Logan raises his eyebrows, takes a half-step backwards and starts to type on the keyboard to fix the error.

"I submitted my thesis," I dare to say, standing at the bench behind them.

Mark turns around. "Good. And now?" he asks, much more friendly than I anticipated.

"I leave for South America tomorrow. Backpacking for a few months."

I don't say that the location as such has been secondary in the decision where to fly to. That I just need a break before embarking on a new stage in my life, though I have no clue what or where this new stage is. As there is at least a month between submission and the PhD defence, I thought this would be the right time to get away from this phase of my life.

"If you think that is the right place to prepare for your defence…"

"I think it is."

"Write me when you are ready and I'll book a date with the examiners."

"Thanks."

I walk to the office and chat for a few minutes with Linn, until Mark and Barry have left the lab and I can say goodbye to Logan.

"I wish I was flying to Buenos Aires tomorrow," Logan says when he enters the office.

He is being funded for four years instead of three. Hence he still has a year to go. "I'll send you a card!"

I lift my backpack and jumper. "Okay, time to leave!" I say.

"Have a good trip, and see you for your viva," Logan says.

"Take care, Ka," Linn adds.

We had already properly said goodbye yesterday evening on my farewell night, so it seemed over the top to make a big fuss about it now. Plus, I would be back… in a few months… I suppose… I really don't want to think about it though… I smile and walk out of the door.

It is just after five when Greg and Felix buzz my doorbell.

"Your movers have arrived," Felix says, smiling at the apocalyptic waste-land that is now my flat.

"This is the stuff I'm bringing," I say, pointing at a large backpack in the middle of the room, surrounded by a few boxes to send to my parents, plus some loose pans that didn't fit in the boxes I have asked Greg to keep until my return.

"What are we going to do with the rest? Second hand shop downstairs?" Greg asks, referring to all the furniture in my flat.

It would have been so easy to just drop off everything in the shop below my flat, and it would have had an appealing tidiness about it. Alas, I had asked if they wanted it, and they politely declined. I couldn't remonstrate; the cheapest furniture IKEA has to offer with three years of wear. My furniture is just awkwardly stacked firewood.

"They don't want it," I say.

"Then we have to make them want it," Felix says.

None of us possesses a car, so getting rid of furniture is a major hassle.

"Yep. They have two entrances, and sort of two rooms, right?" Greg asks.

"Sort of, yes," I confirm.

"So you distract them at one end of the shop while Felix and I dump the stuff at the other end."

"Sounds reasonable."

"We need to hurry though, they're closing soon," Felix says.

I walk downstairs and enter the second hand shop yet another time in just an hour. The middle-aged lady behind the counter obviously recognises me as the mad Dutch girl who thought all her ghastly cheap worn-out furniture was sellable.

"Anything else I can help you with?" she asks curiously, with the same friendly voice she had before.

"Yes. I actually decided against moving and am interested in the table you have over there."

"Right…," she says and, without doubt, looks utterly perplexed but not yet suspicious as such. I told her an hour earlier I'm emigrating to an unknown destination. *She probably just thinks I'm mentally ill.*

"I know, don't ask, it's all very strange," I say, hopefully taking her worry away and leading her to a table as far as possible from the entrance Felix and Greg will use.

The shop spans the ground floor of two large tenements, with a wall and a small door in between, so it is large and sufficiently separated that one can never be sure what is happening on the other side of the shop. Nevertheless, between all the questions I ask her about the butt-ugly granny table and the accompanying chairs, I can hear Felix and Greg dragging my furniture into the shop. There are lots of people bustling about and a wide variety of noises so, thankfully, the lady doesn't seem to register Felix and Greg's contribution to the backdrop. After just over five minutes and enquiring about every silly little detail of the table, I ask her to tell me something about the even uglier table next to it. With little passion, and an increasingly suspicious undertone, she says a few words about the oval table which, presumably, she probably also finds butt ugly. A few sentences later and the suspicion is palpable, and unbearable. *Oh God, what if she thinks we're robbing the place?*

I tell her that I just need to let everything sink in, but will probably fetch one of these two beauties tomorrow. At undoubtedly suspicious speed, I leave the shop through the nearest exit. I walk along the pavement to the other exit and see my sofa, rocking chair, metallic kitchen table and red bench piled up in the shop window.

"We're not finished yet, flower," Greg says walking down with the second red bench in his hand, clearly disappointed to see me.

"She got too suspicious."

"Right. Then we have to wait until they are closed and we place the rest in front of the shop," Felix says.

We walk back upstairs and we all open one of the beers that Greg brought. Less than thirty minutes later we are taking furniture down again.

We place it in front of the shop but it turns out there are plenty of people passing who see a use for my old tat. Most of the stuff we place outside the shop is taken by the time we are back upstairs. *Maybe I could have just left everything there and avoided the suspicious subterfuge?*

"Wow, this is a real gypsy move," Felix says.

Enthusiastically, he runs back upstairs to fetch the next load. It only takes half an hour until it is only my bed and a wardrobe left.

"Where are my pans?" I ask, staring at the pile of stuff in the middle of the room.

"Felix just brought them down," Greg says.

"I did indeed."

"Oh no!" I shriek and run empty-handed down the stairs.

Frantically I look around the pavement, but the pans are gone. My last trace of personal wealth has been taken by a stranger.

"What's wrong?" Felix asks, looking at the stuff on the pavement.

"You donated the best pans ever to a complete stranger!"

Felix looks genuinely sorry. "I'll buy you new ones."

"No, forget it. I don't need pans in South America."

It is after ten in the evening before the whole flat is in a sort of reasonable state. The few boxes with stuff I want to keep are at Greg's place, and it is only me, my backpack and a few dirty towels left. I walk once more to the window sill and look outside. At Gorgie Road. At the City Farm. At the Hearts supporters pub. I am torn by emotions that I know can only be numbed with sleep; happiness to have had my last day at Lab 262, excitement about penguins, Latinos, salsa, wine and all the other things waiting for me, and desperate sadness for all the good people I am leaving behind. There are a few tears running down my cheek. Greg is standing next to me and puts an arm round my shoulder. "It's fine, flower. All will be fine."

We drink another beer, the three of us in the empty flat, and close the front door behind us before midnight. I write a short thank you note to the agent who had let me the flat for three years, tape it to the keys and throw the keys in the mailbox. We head to Felix's shared apartment, drink more and the next thing I remember is the alarm going off at an unseemly hour. *It's time to leave.*

Chapter 42

I follow Prof. Gilton through the chemistry building on auto pilot. *Is my PhD defence really finally happening?...* I have postponed this oral exam as much as anyone can. Under my arm I carry a copy of my thesis, full of dubiously scribbled notes, while my legs move forward unprompted and my hands all but squelch with sweat.

For two months I have revised, from early morning to late evening, all the facts and theories around all the techniques I used during my PhD – every last one of them. Felix has donated hours and hours of his free evenings and weekends to explain to me every bit of chemistry in my thesis right down to the last dull detail. He turns out to be a born teacher; he has instructed me better than any Russian spy might be briefed before infiltrating the CIA. Just yesterday, it seemed I knew *everything*, but now my head feels empty.

"How are you, Karin?" Prof. Gilton asks, opening a door to the new part of the building.

"Considering the circumstances I am quite fine."

"You said hello to Mark?"

"No, I avoided him."

With a supportive smile, Prof. Gilton turns to face me. "Say hi afterwards, okay?"

"Okay."

We enter a small room at the back of the chemistry building, where I have never actually been before. Prof. Gilton tapes a note to the door: "Viva in progress. Please do not disturb!"

© Springer International Publishing AG 2017
K. Bodewits, *You Must Be Very Intelligent*,
DOI 10.1007/978-3-319-59321-0_42

A grey-haired guy, initially sitting with his back to the entrance, springs up to introduce himself.

"Hi Karin, Professor Walt Green, University of Aberdeen, pleasure to meet you."

He speaks with a soft voice and offers me a cold, limp-fish hand. He is attired in a pair of dark blue trousers which are not only too long but also far too wide. The curious garment is held to his body by a brown belt. The short-sleeved white shirt with grey stripes is also at least two sizes too large for this man of normal size and weight. *What is that about? Do they not do small sizes up North?* He's wearing a pair of grey eighties glasses that he probably bought during his college days – I can't imagine anyone still selling them, not even in a charity shop. Something in his careful smile tells me he knows I gave him the once over and judged his clothing. *Woops.*

I smile back at him. *Oh dear, he looks vulnerable.* Suddenly I picture him as a schoolboy being relieved of his lunch box by bullies, daily, always the last boy picked to join a sports team. We haven't even started and I'm already feeling sorry for my examiner. *How weird.* And I get the impression he is somehow more afraid of this exam than I am. *Who the hell brought this loser here?!*

I sit on a chair on the other side of the table. Despite Prof. Green being much less scary than I had nervously envisaged, my anxiety levels remain stratospheric. This man will decide if I pass or not. Very soon…

"Who do we have here?" Prof. Green asks in a friendly voice. *Who could I be? Has he been lured into this at the last minute and he has not actually seen my thesis?*

"Eh… I am Karin…"

"I know your name. But I would like to learn a bit about you before we start. What drives you?" *WTF dude! I am here to be grilled about my thesis. Since when was I ever asked what drives me, the person carrying around four limbs, two of which can hold a pipette? I thought that was all that mattered here.*

I stare at Prof. Green, not knowing what to say. For several seconds I grapple to find an authentic answer. I don't have one. Apart from passing this exam, all that drives me these days is an old hobby I recently picked back up – painting nudes, always women and ideally prostitutes. But that seems a rather obscure thing to mention during my PhD viva. In the past, love of science drove me, but today I have no clue what motivates me or why I am here, going through this farce to get my hollowed-out PhD.

"I am not sure… I guess at the moment I'm mainly driven by wanting to have this exam out of the way," my voice is trembling.

"Do you have a job?"

"No, not yet."

"Karin has been travelling the last few months," Prof. Gilton says to Prof. Green, trying to tweak the conversation in a different direction.

"Where have you been?"

"Argentina, Bolivia and Colombia," I answer.

At least I could answer that. Despite my PhD defence hanging over me like the sword of Damocles, it had been a great trip.

"How was Colombia?" Prof. Green asks, truly curious.

"I learned to paraglide there."

"Paraglide?"

"Yes."

"Something you always wanted to do?"

"Not really, but… I took the opportunity when it came along."

I don't mention that I took a paragliding course, lasting several weeks, on a mountainside in Colombia with an instructor so stiff-dazed full of cocaine that he wouldn't have noticed had one of his students crashed into a pylon right in front of his nose. Owing to all-round incompetence, I flew mere inches above electricity cables and almost landed in a river due to incomprehensible instructions from a voice destroyed by alcohol and smoking. I watched peers make full-body tree-landings, crying for help with feet dangling inches above ground. I don't mention any of it, though it occurred to me at the time that three years before I wouldn't have signed up for these evidently suicidal exercises. Briefly I wondered if my time at the University of Edinburgh has rendered me nihilistic, feeling worthless, nothing to lose…

Prof. Green looks at me and smiles. "So you like to put your feet above ground."

"Actually no, I'm afraid of heights."

"You did a paragliding course whilst being afraid of heights?" Prof. Green asks, in disbelief.

"It was awful," I say, laughing uncomfortably.

By now I am wondering if this is the only external examiner Mark could persuade to come to Edinburgh to do my viva. The gilded robes like Homer Simpson and Professor Johnson could probably persuade any self-respecting researcher to fly in and conduct an exam – if only to increase the chance of future collaboration. I guess Mark can only find examiners who have never heard of him before or are delighted to be asked by just about anyone.

Prof. Green is peering at me. I realise he honestly wants to know about the girl who wrote the thesis in front of him.

"What kind of job do you want to do next?"

"I don't know really."

Four years ago, I had dreamt of becoming an excellent scientist. Now I don't even have the confidence to be sure I would make a good toilet cleaner.

After finishing, Lucy went off to teach at a school in Senegal; Babette was apparently planting trees in Canada to avoid any human contact; Erico was working on an oil rig close to Greenland; Quinn bought a dog and followed his wife to the US; Hanna moved with her partner to the south of the UK; Diet Coke Girl chilled at her parents' holiday house in Cyprus; about Linda I'm not sure, as she has never been mentioned again by anyone. But whatever they did and wherever they went, eventually, Stockholm Syndrome claimed most of them and they returned to science. Erico has been the only one to remain outside of academia, on the rig, but he studied something on the side. Maybe I will have a similar fate? I don't know. I so badly want to be honest and authentic but I don't have the faintest idea what kind of job I want next. I only know for sure I don't want this one.

"Do you feel you learned enough during your PhD to prepare you for the next stage in your career?" *Learned enough? Had I learned anything?*

Profs. Gilton and Green are scrutinising me, awaiting a reaction. I take a deep breath. I am trying to overpower tears welling up in my eyes. How can I tell the truth? That I have "learned" as much as a penguin learns when he eats yet another fish? That I had "learned" that I am grossly imperfect, nay downright deficient, and oh-so hopelessly weak?

"I guess there is always more to learn," I manage to say as if the question hadn't nearly made me weep.

I am desperately hoping we now move on. "Prof. Green is just asking you these questions to calm you down, Karin," Prof. Gilton says.

"I know, sorry, I'm just not prepared for them."

Both men open a copy of my thesis. Prof. Green's is scribbled full with notes, and we are only on the acknowledgement page.

"Before we start, I would just like to know if you are a biologist or a chemist by training," he says.

"A biologist!" I reply much quicker than I had intended.

Within a fraction of a second I have awful flashbacks to my first year viva. Despite Felix's chemistry boot camp, biology is my much preferred métier.

"Let's go through it page by page," says Prof Green.

I look at the 250 pages in front of me, and then at all the notes he has made in the margins of his copy. I look despairingly at Prof. Gilton. He avoids my eye, but he too looks alarmed by this barking mad suggestion.

"Eh… sure," I say.

He patiently flicks through the first thirty pages of the introduction. "It's very good," he mumbles.

He doesn't look up from the pages and he doesn't ask me any questions. *Oh, if we're going through every single page by looking at it without comment, just checking it exists… great! I should pass no problem because I'm a dab hand when it comes to staring at pages in silence.*

At length he points at a chemical formula, and redraws it for himself. In all seriousness, he informs me that one of the dozens of hydrogen atoms is pointing in the wrong direction. *A-ha! You're a member of the hydrogen atom mafia, which Felix warned me about! You're one of those many chemists who get their knickers in a knot about the direction of every single atom.*

He continues silently flicking through and arrives at page sixty, whereupon he asks me to draw four different types of sugar molecule. I could do so blindfolded. Prof. Gilton dutifully checks the piece of paper I drew the first two on, and that is as much interest as he can fake in this bewilderingly pointless exercise.

"Keep on, I'll be right back," he says and leaves the room, carefully closing the door behind him lest any noise distract from the dull delineation of a dull sugar molecule.

Prof. Green looks worried and confused about Prof. Gilton leaving the exam. I guess university policy stipulates that there should be at least two people present during a PhD viva – not only to help out in cases of disagreement but also to prevent accusations of sexual interactions between examiner and student. Alex and Chris had told me in the pub that they had been offered blow jobs by a female undergrad in return for a higher grade. "Did you take it?" I asked curiously, knowing that if they did accept it they probably would not have told the story to Lucy and me.

"Of course not! In my case she was gorgeous, but it would simply be wrong," Alex had declared.

"The girl who offered it to me was just too ugly," Chris had added, ironically.

We had asked them if this was a common occurrence. They both said it only happened to them once, but they had heard whispers and rumours from other male colleagues. I had wondered how that might play out? How could the girl hold him to the deal? How could she complain to the Faculty about having sex for a better grade which was then not forthcoming? *"I'm going to sue! My sexual technique is worth a First! I'll prove it in court!"*… And could he say: *"What? Just one blow job?! Excuse me, my good lady, but the going rate is several nights of depraved and torrid sex!"* The whole dishonourable business would require an enormous degree of trust between shifty manipulators.

For now, while we await the return of Prof. Gilton, I regard Prof. Green on the other side of the table, sitting with rounded, hanging shoulders in his outsized clothes, and I wonder if there are any students desperate enough to give him a blow job in return for a pass mark. *Would I be willing to do that?*

I try to imagine Prof. Green in skinny jeans, nice T-shirt and hip glasses, but I conclude that he is irredeemably unattractive or, at the very kindest, really not my type in a million lifetimes. I feel disturbed by that thought; so sex is not an option purely because he is too unattractive? *What if he were a guy I could fancy? What if Alex was conducting the exam?* A momentary feeling of relief mingled with dignity passes through me as I realise it would never be an option, I would never offer my body in return for a full degree. *Phew, my moral core has not entirely atrophied...*

Cruelly and inaccurately – and in a manner I will regret afterwards – I wonder if Prof. Green is of a particular type of academic; of that stripe which will inevitably be reduced to preying upon impressionable undergrads. As I said before, in my more cynical moments I sometimes wonder if academia is maintained partly to give males without sex appeal some chance of sex at some point in their otherwise barren love lives...

I fear I may have been staring at Prof. Green for too long with a subtle – or maybe not subtle enough – expression of distaste on my face. He is raising his eyebrows, looking at me as if he can read my thoughts, seemingly asking me to please focus on the stupid drawing of the stupid sugar molecules again.

"You and Prof. Gilton do not know each other, do you?" I ask quickly.

"No, we don't."

"He'll be back. He probably just went for a smoke."

Prof. Green looks at me as if I just told him that Gilton has gone to collect his weekly supply of Rohypnol. Briefly I consider mentioning that he is probably downing a few shots of Jägermeister as well. But I concentrate on drawing the molecules instead.

Half an hour into the exam, Prof. Gilton returns to fill the room with the odour of someone who just smoked a cigar in a closed office room. Prof. Green asks me questions about the techniques I used, my results and how I came to certain conclusions. All the questions I am asked – none of which are about me as a person – I am able to answer. Shortly afterwards the question-answer scenario gives way to a normal grown-up conversation between adults about research. I am not nervous anymore. The sweat in my palms has evaporated. I am confident I will pass.

Just after the hour mark Prof. Gilton opts for another break. Again Prof. Green looks distressed, but not to the same extent as the first time he had been alone with me. Again Prof. Gilton returns smelling of smoke, but this time he had kindly brought coffee as well. We were all rubbing along fine.

Almost two hours in, Prof. Green is approaching the end of the thesis.

"So you published two papers, and might submit the third soon, right?" he asks looking at the Angewandte-Chemie-Christmas-present-co-authorship and the silly LpxC paper in the appendix.

"I don't know if the third paper is going to be submitted soon."

"Do you still need more results?"

"Not really, no. But it has been on Mark's desk for months now, so I'm not sure if it's ever going to be published."

I omit the story about the authorship argument I had with Barry. I have no clue what happened, but I suspect Mark and Prof. Gilton have talked about the incident. And I presume Mark has decided against publishing it altogether. In other words, a possible side effect of this academic over-competitiveness is that people suffering from a deadly disease will not benefit from the small chance that this research might have one day brought forth a life-saving treatment. But I can't ask Mark about it anymore. Even though publications are the be-all-and-end-all of academic achievement, allegedly, I would rather leave my PhD with one publication less than dredge up stuff that cost me so many sleepless nights.

Prof. Green does not enquire further and indicates to Prof. Gilton that he has now finished his part of the exam.

"Okay, Karin," Prof Gilton says. "I ask you to leave the room and wait outside for five minutes or so."

I stand and notice that I have left a most undignified sweat stain on the plastic chair. Quickly I push it under the table, smile at both gentlemen and walk out of the room. I can hear their voices from the corridor, but I can't understand what they say. In contrast to my first year viva, where I had been a nervous wreck during the wait outside of the room, I feel totally at ease. There was nothing to hint that I would not pass. I had answered everything they had asked. And they did not say anything about my profound lack of interesting results, or voice misgivings to suggest I might have to repeat or redo anything.

I lean against the wall and my head starts to throb painfully. It seems my body, finally able to relax after hours of stress, can now reveal that it feels terrible after this prolonged suppression of normal operation. The headache is followed by cramps in my empty stomach and a huge heaviness in my legs. I close my eyes for a moment and take a few deep breaths. I feel light-headed as if I just stood up after a too-warm bath. I have an out-of-body moment and see myself standing in front of the door where the PhD defence had taken place. I see a disillusioned and defeated doctor-to-be, without any future plans, to whom a degree from a famous university means nothing anymore. It is the same girl who started her PhD almost four years ago feeling ambitious and energetic, manically driven with the desire to become a scientist. Now she is just drained and bored, and not very healthy.

"Karin?"

I open my eyes. Prof. Gilton's yellow tobacco fingers are patting my shoulder.

"You can come back inside," he says quietly.

I follow him, and sit down on the same chair.

"Congratulations," Prof. Green says, with an excitement I do not feel, and shakes my hand.

"You passed, very well done!" Prof. Gilton adds.

I smile but I don't know what to say. I am not sure if I should be proud or, weirdly, feel humiliated.

"Just four hydrogen atoms to correct and that's you done," Prof. Green says.

"Is that all?" I ask, in disbelief.

"Yes, that's all."

"Wow! Thank you very much."

"It has been a pleasure discussing your PhD with you, Karin," says Prof. Green. "You have some good ideas in your head."

I press my lips together and glance downwards, not knowing how to handle the compliment.

"Can I ask you one more personal question?"

"Sure."

"What did you like most about your PhD?"

"Writing my thesis," I say.

I had put a lot of time into the writing, especially the sixty page introduction; a review of the literature ending with where I had hoped to contribute to the scientific world. I had loved digging through literature, finding interesting facts that I could add to make my text more engaging. I had spent many long and happy hours at my desk in the living room reading and writing rather extraneous stuff.

"I suspected you liked that. It is very well written. Let's be honest with you: your research results are okay but not flabbergasting. But, I actually very much enjoyed reading your thesis. *Do* something with that." *I will.*

I walk out of the bathroom and scroll through my contacts for Felix's number.

"And?" he says without even a "Hi."

"Your boot camp got me through," I say, finally feeling excited.

"Cool! Any corrections to do?"

I look up and down the hallway to check for Prof. Green. "He turned out to be one of those hydrogen atom freaks you warned me about. So I need to change the direction of a few in some molecules."

Felix laughs. "That's all?"

"Yup."

"Very good."

"I need to pass the lab and will be in KB House in half an hour for an urgent beer."

"I'm already there, having coffee with Greg, waiting for you."

Nervously I let my mobile slip into my bag and press my upper body against the heavy door of Lab 262 to open it. Carefully I walk in, as if placing my feet on forbidden ground. I look around but there's no one working at the benches. Nothing has changed in the last nine months, since I was last in this room. Profoundly reassuring. The huge authentic autoclave stands in the same position. The note "not in use" is tacked on its door. The benches are messy and old machines are stored under it. I walk to the office. My desk contains the old computer with the broken fan, which I had gotten from my parents almost two years into my PhD. I had given it to the two new PhD students who started the day after I left.

"Good to see you, Ka!" Logan says.

"Good to see you, too!" I say, wishing I could hug the guy I worked with for a few years. But I suppress the urge to wrap my arms around him. I know too well he isn't the type for it.

"Did you pass?"

"I did."

"Congratulations!"

"Thanks."

"How did it go?"

"It was kind of strange. They were both very friendly. They asked me a few questions and then the whole thing just went into, well, an interesting conversation about science. And the external examiner was a bit like a… a pussy. But nice."

"Don't say that, Ka," Logan says, smiling.

"Well, he was! Anyway, Prof. Gilton went out twice during the exam for a smoke. Can you believe it?"

"No way," Linn says, just entering the office.

"Way."

Logan smiles. "Good old Gilton."

"Have you seen Mark yet? He complained this morning that you didn't go to his office to say you'd arrived."

"Not yet, no. I was too nervous to handle Mark on top."

"How was your trip?"

"Very good. Nice weather, nice food, hot Latinos…"

"No details, please!" Logan says.

"You don't want to know how they swing their hips?"

"No!"

"But I want to know…" Linn says, playfully curious.

"How are things here?"

"Same old shit."

"Kate is babysitting Mark's baby, once a week," Linn whispers.

Kate is one of the new PhD students, who I have only seen once; a beautiful young English girl with cascading brown hair.

"Does she want to do that?" I ask, feeling terribly sorry for her already.

"I don't think so."

"Does she get paid for it?"

"I don't know."

"Poor girl."

We hear the door of the lab open, and the key chain we are all so familiar with. All three of us fall silent. Linn and Logan turn to their desks and pretend to read. I'm just standing there like a lost animal in an alien environment, waiting for Mark to enter the office. We hear him talking enthusiastically in the lab.

"Ah, someone's here for an interview," Logan says with a soft voice.

Mark shows the guy round. "Don't worry about the mess… in the process of tidying it all up…"

I walk into the lab as it would be just too weird – and obviously cowardly – to stay silent in the office.

"Hi," I say not overly enthusiastic.

Mark looks towards me.

"I heard you passed your exam, congratulations," Mark says, trying to sound relaxed, but the tension is electric.

"This is Karin, she's been working on the LPS project, wrote an excellent thesis, just passed her viva," he says, in a voice that smacks of forced levity, to the young, black-haired guy standing next to him.

"I didn't know you read it," I say, with a sarcastic smile on my face.

Mark's eyes are shooting fire at me but he laughs my barb away for the benefit of the prospective employee.

"Hi, a pleasure to meet you," I say, extending my hand.

"Sebastian," the guy says in a lively, youthful way.

"There's a bottle of champagne in the cold room," Mark says. "I won't have time to drink it with you so just help yourself."

His eyes are still shooting fire at me but he strains, successfully, to keep the tone friendly. He walks to the office where Linn and Logan are still pretending to work behind their computer. Sebastian follows closely behind.

"This is Sebastian from Spain," says Mark brightly. "He will be giving a talk in an hour. You are all supposed to attend."

"Hi," I hear Linn and Logan say at exactly the same time.

"Where are Kate, Patrick and Barry?" Mark enquires.

"Went for lunch, back any time now," Logan says.

"Tell them that I expect them at the presentation as well."

"Sure."

Mark attempts a few jokes with Linn and Logan, who laugh along politely enough, and then heads out of the lab taking Sebastian with him.

As soon as the door closes, I fetch the bottle from the cold room.

"Shall we drink it?" I ask Linn and Logan.

"I could use a glass," Linn says.

Logan nods. "Me too."

We empty the whole bottle within ten minutes while gossiping about the lab. When the glasses are empty I pack my bag, ready for the next drink stop: KB House. Logan and Linn agree to hook up with us around five, when the Polish Vodka Bar is on the itinerary.

I close the door behind me and walk across the building. I pause in front of Mark's office and hear him chatting away in enthusiastic fashion about the research projects he has in mind and the great collaborations he's got going. That had been me four years ago, listening to his fiction about the Promised Land.

"Don't take it!" I whisper, smiling sadly at the closed door.

Of course Sebastian would not believe it if I were really to warn him; just as future students won't believe it if Sebastian tries to warn them… And so it goes on, unchecked, oppressive, soul-gnawing, enervating little slave empires run by tin-pot paranoiacs preying on gullible, hopeful, dreamy youth. It is too sad to countenance. And noble science deployed as a guise is perhaps the saddest thing about it.

Epilogue

Dear Mark,

Thank you very much for reaching out to me. I am currently not reading my emails, and I am not sure I will ever pick up the habit of doing so again. However, don't feel too disappointed. This email is being stored in the cloud and added to a long queue of other emails that are being sorted and deleted randomly. The average time of processing is, based on the number of emails waiting today (Total No. = 2), approximately 17 weeks.

I have a mobile device on which you could easily reach me. The number, which is also routed through my landline, consists of the following digits: 0 (5x), 1 (1x), 2 (1x), 3 (1x), 4 (2x), 5 (1x), 7 (3x), 9 (1x). For privacy reasons I cannot provide the correct order by email.

Be patient, stay strong.

Your ex-PhD student,

Karin

© Springer International Publishing AG 2017
K. Bodewits, *You Must Be Very Intelligent*,
DOI 10.1007/978-3-319-59321-0